OBSERVATIONS ON GOLDEN EAGLES IN SCOTLAND

ADAM WATSON
&
STUART RAE

ISBN-13: 978-0-88839-030-1 [trade edition softcover]
ISBN-13: 978-0-88839-173-5 [trade edition hardcover]

Cataloging in Publication Data

Library and Archives Canada Cataloguing in Publication

Watson, Adam, 1930-, author
Observations of golden eagles in Scotland / Adam Watson
& Stuart Rae.

Includes bibliographical references.
ISBN 978-0-88839-030-1 (softcover)
ISBN 978-0-88839-173-5 (hardcover)

1. Eagles--Scotland. I. Rae, Stuart, 1956-, author II. Title.

QL696.F32W38 2018   598.9'4209411   C2018-902019-9

Printed in the USA

PRODUCTION & DESIGN: M. Lamont
TECHNICAL EDITOR: D. ELLIS

FRONT COVER: Two Golden Eagle chicks in an eyrie in a Scots Pine in the north-east study area. Photo by Stuart Rae.
BACK COVER: Late-lying snow in the high corries of Cairn Toul and Braeriach, in the Cairngorms. Photo by Stuart Rae.

hancock

house

Published simultaneously in Canada and the United States by

HANCOCK HOUSE PUBLISHERS LTD.
19313 Zero Avenue, Surrey, B.C. Canada V3Z 9R9
(604) 538-1114   Fax (604) 538-2262

HANCOCK HOUSE PUBLISHERS
#104-4550 Birch Bay-Lynden Rd, Blaine, WA U.S.A. 98230-9436
(800) 938-1114   Fax (800) 983-2262

www.hancockhouse.com   sales@hancockhouse.com

OBSERVATIONS ON GOLDEN EAGLES IN SCOTLAND

ADAM WATSON
&
STUART RAE

TABLE OF CONTENTS

A young golden eagle chick, in its second week, lifts its head to look around while its sibling lies sleeping. Pine needles make a surprisingly soft bed for these birds in the crown of a Scots pine.

An eaglet stands defiant after being measured and ringed. It has a British Trust for Ornithology numbered metal ring on the right and an individually colour and alpha-numeric-coded one on the left, which is easier to read when seen in the field. It is on the ground because it had been lowered from its tree eyrie to be processed; that is safer for the bird.

This volume is dedicated to the memory of Leslie Brown in appreciation of his early studies on golden eagles in Scotland and for his inspiration to the authors to continue such work. The early data on eagles from the north-west Highlands which we have used in this volume originated in studies by him and Adam Watson in the 1950s and '60s.

Eagle country in north-west Sutherland. A well-acquainted view of Leslie Brown and Adam Watson in the 1950s and '60s, and little-changed in the 1980s and 2000s when familiar to Stuart Rae and Derek Spencer.

Foreword

Majestic in flight, with a grip out of proportion to its size, denizen of wild places, elusive and retiring, yet stunningly beautiful when seen near at hand; all of these phrases hint at the allure of this great bird. The golden eagle holds pride of place in the culture and natural heritage of Scotland. A day in the hills is enlivened by just a fleeting glimpse, and many naturalists have decades-old, but still vivid, memories of their first, close encounter with this formidable raptor. Seeing but a hint of its predatory prowess, watching territorial displays, or merely gazing at a soaring bird are all part of the pleasure of studying the eagle and trying to learn its secrets.

The research and conservation efforts on golden eagles in Scotland, as reported in this exceptional book by Adam Watson and Stuart Rae, are extraordinary. Golden eagles have been studied in Scotland for over a century. H.B. MacPherson's *The Home-Life of a Golden Eagle* (1909) was the first formal monograph on the bird, and was ahead of its time, with photographs of a nesting pair on a cliff in the Highlands. Captain C.W.R. 'Chaz' Knight lectured widely in the US, Canada and Britain, accompanied by his tame eagle 'Mr Ramshaw', and both appeared in the film *I Know Where I'm Going* (1945). Seton Gordon was the first naturalist dedicated to studying eagles in Scotland. His *Days with the Golden Eagle* (1927) described early observations in the field and notes made from hides near nests. It is still a classic and obligatory reading for students of eagle behaviour. Ably assisted in the field by his wife Audrey, he presented further important ideas in *The Golden Eagle: King of Birds* (1955).

Since the 1950s, we have seen the emergence of an outstanding cadre of field naturalists studying golden eagles and other raptors in Scotland. Charlie Palmar continued with impressive photography and detailed observations on numbers, breeding success and food taken to nests in Argyll, and Pat Sandeman wrote a fine short paper on numbers and breeding success in relation to land use in east and central Perthshire (Scottish Naturalist 1957). In 1955, Leslie Brown, with the help of Charlie Palmar and Adam Watson, compiled an account of all pairs and nest sites then known to them, a useful baseline for decades later. Information on eagles was now amassing, and Leslie Brown's *British Birds of Prey* (1976) provided a large amount of detail. Desmond Nethersole-Thompson's important observations on conservation were included in T*he Cairngorms: Their Natural History and Scenery* (written with Adam Watson, 1974, 1981); Desmond was a great advocate of long-term studies to unravel the mechanisms controlling population changes and range occupancy. David Stephen (one of the earlier popular writers on Highland wildlife) and Jim Lockie made valuable studies of the diet of eagles in remote regions. Derek

Ratcliffe summarised his and others' observations in *Bird life of Mountain and Upland* (1990), including some of his own early and ground-breaking work with Jim on the insidious role of pesticides in eggshell thinning. Roy Dennis collated and coordinated much of the monitoring work, and has promoted an eagle-friendly management attitude to the Highlands and Islands. Meanwhile, in the background solid long-term work has been done by Robert (Skitts) Rae, Liz Macdonald, Keith Brockie, Mick Marquiss, Stuart Benn, Ken Crane, Kate Nellist, Derek Spencer, and many others who have become associated with the Scottish Raptor Study Group.

In 1997 (updated in 2010), Jeff Watson brought Scotland to the global forefront of eagle research with his definitive monograph *The Golden Eagle* (brilliantly illustrated by Keith Brockie). Since then, most written reports on eagles in Scotland have appeared in scientific journals. Based on several research papers, *A Conservation Framework for the Golden Eagle: implications for the conservation and management of golden eagles in Scotland* (Scottish Natural Heritage, 2008), by Phil Whitfield, Alan Fielding, Dave McLeod and Paul Haworth, made sombre reading regarding the impacts of persecution and heavy grazing on golden eagles. However, a national survey in 2015 estimated just over 500 breeding pairs, thus revealing recent improvements in eagle conservation in some parts of Scotland. Pioneering eagle work in Scotland was brought to the fore in David Ellis's world-wide anthology *Enter the Realm of the Golden Eagle* (2013). The knowledge gained from studies of golden eagles in Scotland has indeed come a long way since the pioneers, and fundamentally influenced eagle research and conservation around the globe.

That brings us back to this book and its authors—the energetic and resourceful Stuart Rae, who wrote *Eagle Days* (2012), and Adam Watson, who is a living legend for his research, writing (including 27 books) and campaigning. *The Cairngorms of Scotland* (1998) was their first book together, and now we have this soon-to-be classic. Spending 47 and 75 years, respectively, studying eagles, Stuart and Adam are eminently qualified to write about their own work and also to overview the long-term work of others.

A few important themes form the basis of this work. First, it is only through long-term, painstaking field work, and careful analysis, that we can truly understand eagle demographics and ecology. Second, persecution has for centuries limited the numbers and distribution of golden eagles, and, to a degree, it still does. The Victorian era's practice of extirpating large predators is supposedly over, yet we continue to see evidence of poisoning and shooting of adult and young eagles. Third, land use affects eagle occupancy and food supply. High densities of sheep and deer can give rise to heavy grazing impacts on vegetation and soils, and of course carrion as eagle food. This, and 'muirburn', conifer afforestation and wind turbines all have influences, and with regional variations, we see differences in density and breeding success of eagles.

The golden eagle is so widespread across Europe, Asia, northern Africa and North America that regional populations will surely be differentially affected by changes to their local environments. Some might not be favourable, but from the evidence presented here on eagle demography in Scotland, there is clearly much that can be done to secure a better future for the birds. By delving into the underlying effects of land use, habitat changes, disturbance, persecution, nest sites, and food quality and availability, the authors uniquely provide us with a fascinating diagnosis of the state of Scotland's golden eagles and their study, and a prognosis for their future.

In Scotland, and across the golden eagle's vast range, this great bird is a natural treasure. Too often it is sacrificed to an unseeing public when, with some encouragement, local people and visitors would befriend and defend the eagle. In the South of Scotland there are the beginnings of a major project to restore eagle numbers; and in several regions there is local pride in population increases and some evidence that golden eagles are faring better now than they have for decades. On the Isle of Harris there is an encouraging drive to build tourism around eagle watching. Now is the time to embrace the king of birds for, in its fortunes, we reveal our own commitment to nature. People living with a healthy eagle population are to be congratulated for their conservation ethic, for, in saving the eagle, they save wild lands, and a host of other wild things.

The gravity of each human endeavour is judged on the importance of the cause, the difficulty of the task, the magnitude of effort, and the ingenuity of the solution. This book deserves rich accolades for the massive, persistent effort of the authors and their colleagues. In the annals of single species studies of raptors there is no comparable project, and no comparable volume, to what Adam Watson and Stuart Rae present here.

David H. Ellis, Oracle, Arizona
Des B.A. Thompson, Edinburgh

January 2017

Introduction

This book brings together long-held, up-to-date unpublished information on golden eagles *Aquila chrysaetos* in north-east and north-west Scotland. Some of the contents are descriptive in style due to the nature of the studies, which includes historical information secured from predecessors with intimate knowledge of eagles in the areas of interest, such as deer-stalkers, gamekeepers and naturalists. However, most of the data came from personal studies by Adam Watson (AW) and Stuart Rae (SR), and have been analysed afresh or in context with data from their previously published work. Because most of their comparable data are relevant to north-east and north-west Scotland, those are the study areas emphasised in the following chapters. The aim of this volume is to collate the authors' combined knowledge of golden eagles, share that knowledge with others, and interpret our findings with a broad comprehension of how golden eagles have fared, are faring and will probably fare in the future in Scotland.

There has been a long history of studying golden eagles in north-east Scotland since the pioneering work of Seton Gordon at the end of the 1800s. AW began his own studies in 1943 and SR in the mid 1970s. The main area of all these authors' studies was upper Deeside, around and beyond Braemar, Aberdeenshire. Because AW had a comprehensive knowledge of all the eagles breeding in that area by the 1950s, he and Leslie Brown used it as one of their study areas for their innovative study of golden eagles and their food supply in 1957–60 (Brown & Watson 1964). That work set a baseline for methods and interpretation of how golden eagles live, which has since been continually referred to by people subsequently studying eagles worldwide.

One of the other study areas used in 1957–60 by Brown & Watson (1964) was in the far north-west, and theirs was the first detailed study of golden eagles in north-west Sutherland. Brown (1969) also checked breeding density and success in the same area during 1967, and extended the study area by three more home ranges not previously identified. Bernard Hendy noted the breeding success of several golden eagle pairs in the area between 1962 and 1989.

In 1982–85, SR monitored all the potential breeding sites in the area and assessed the availability of potential food for eagles as part of a study of several Scottish regions that compared abundance and breeding success between regions in the context of different land use (Watson *et al.* 1987, 1992, Watson 1997). The study area that Watson *et al.* selected in Sutherland was generally the same as that used in the previous studies, encompassing all their home ranges, but larger, with a further three home ranges. SR then returned to the Sutherland study area in 2012 to help

Derek Spencer, who has monitored most of the home ranges since 2011. The study of golden eagles in the 1980s led by Jeff Watson (Watson *et al.* 1987, 1992) used broadly similar methods and re-examined parts of Brown & Watson's study areas. SR was Jeff's assistant in that study, although they seldom worked in the same place but covered separate areas. They studied 144 golden eagle pairs, numerous sites occupied by single birds and many vacant home ranges— about a third of the Scottish population. However, they made no direct comparisons of their findings on the status of eagles with those of Brown & Watson where they had replicate study areas. This gap has now been filled in the present book for the north-west Sutherland study area, where SR did the fieldwork and the data allowed comparisons between the two periods.

Watson *et al.* (1992) also used a study area (their area 4) that overlapped Brown & Watson's study area 1 in north-east Scotland. However, because the areas were not directly comparable and AW held a long run of data on eagles and potential food abundance for the area, there was no re-analysis of Watson *et al.*'s data in comparison with Brown & Watson's data. Rather, in the current book, for studying changes of golden eagle density and breeding success in relation to land use over a long number of years in north-east Scotland, we considered it more applicable to use the long run of continuous data than compare two sets of data with a gap of more than 20 years.

The chapters in this book are laid out as a set of papers, grouped to fit the two study areas, with the first part dealing with golden eagles in north-east Scotland. Photographic sections within chapters illustrate the eagles' habitats, nests, young, food and the work involved in their study. The first chapter puts the historical knowledge of golden eagles in the area into perspective by bringing together many, often brief comments and recollections, and compares the whole with the current and possible future effects of human land use on golden eagles. The second chapter describes eagle nest sites, using information gleaned from 399 known eyries in that one area. Two other chapters discuss the demography and potential food supply of eagles, based on a unique run of more than 100 years of data.

The main study area in the north-east was as described by Brown & Watson (1964) as their ecological Type 1, eastern Highland deer forest (approx. 68,000 ha). The area is typified as having a relatively dry climate, and an abundant wild food supply. It ranges from 200 to 1300 m in elevation with fairly gentle slopes and extensive summit plateaux, freely drained with rather low precipitation at 75 cm in the east and 150 cm in the west. The lower slopes are mostly covered by heather *Calluna vulgaris* with grass on the valley bottoms. Some of the valleys are fairly well wooded with birch *Betula* sp. and Scots pine *Pinus sylvestris,* and plantings of exotic conifers. There are red deer *Cervus elaphus* in all unenclosed areas; the lower slopes support many red grouse *Lagopus lagopus scoticus* and mountain hares *Lepus timidus*; the high hills carry numerous rock ptarmigan *Lagopus muta*; and there are rabbits *Oryctolagus cuniculus* in the valleys. There are only small numbers of sheep on the unenclosed ground. This area now lies within the designated zone 11, Cairngorms massif, of the Biogeographic Zones of Scotland, termed Natural Heritage Zones (NHZ) by Scottish Natural Heritage (SNH 1998). The environment is similar to that in 1957–60 and 1982–85, except now there is extensive use of the area by visitors pursuing outdoor activities all year (SNH 2001), there are fewer red deer and tree plantings, and more tree regeneration.

The second part of this volume describes similar demographic details of golden eagles in relation to food supply, land use and disturbance in north-west Sutherland as was done for those in the north-east. This study was based on Brown & Watson's area Type IV (approx. 90,000 ha), typified by a moderately wet climate, steep hills, low moorland and scarce wild food. The area lies in two main zones, approximately half in the east and half in the west. The two parts of the study area are now representative of two Natural Heritage Zones, the western part being in the North West Seaboard, zone 4, and the eastern part in The Peatlands, zone 5 (SNH 1998). The eastern part features isolated hills rising to 900m above peaty moorland, and the western part consists of more conjoined hills of a similar height falling to sea-level at the coast with less flat or undulating terrain (Brown & Watson 1964). The annual rainfall ranges 150–300 mm but varies greatly between the low-lying coastal area and among the high hills. Arctic-alpine vegetation descends to about 500m with more grasses and sedges in the west and more heather in the east on the lower slopes. There is very little woodland and all eagle eyries are on cliffs. Sheep are grazed outside the few fields all year round, mainly on the lower slopes.

In the 1950s–1970s, sheep grazing, deer stalking and salmon fishing were the main land uses, with a little grouse shooting on some eastern moors. Visitors are also attracted to the area for sightseeing, hillwalking, climbing and wildlife observation (SNH 2001), and there has been an increase in tourism to the area since the 1970s. The numbers of sheep held on the ground peaked in the 1950s and 1990s, then fell, especially after 2005. This was similar to that in other areas of hill farming following changes in Common Agricultural Policy payment schemes (RESAS 2015). Red deer and domestic sheep are the main large herbivores in the area, and their carrion is food for eagles. Other potential food species are rock ptarmigan on the summits and ridges; red grouse on the heather-rich moors; mountain hares on both; rabbits in some lower, grassy areas; and other less abundant or smaller species which eagles can take.

The precise locations of the study areas were not given by Brown & Watson (1964) for the sake of the eagles' protection, because egg-collection was a serious problem at the time. To clarify this: the study area IV, Sutherland lay west of the road from Lairg to Tongue, and mostly north-east of the road from Laxford Bridge to Lairg. The area used by Watson *et al.* extended south to Kylsku and Loch Shin. Because there are no data for any one comparable set of home ranges in north-west

Sutherland for all years of study, the results are presented in two papers. Chapter 6 compares the home-range occupancy and breeding success of golden eagles in north-west Sutherland between 1957–60, 1967 and 1982–85 using a set of comparable data from 16 home ranges over these three short study periods. The next chapter documents the decline of home-range occupancy and breeding success of golden eagles from a smaller sample of home ranges in the same region over a longer time, 57 years. The following chapter, relevant to these, discusses the possibility that a decrease in eagle breeding density and success was related to a decrease in abundance of food.

The final chapter gives a brief overview of the studies' findings and discusses how these results could be used to guide further research on golden eagles, particularly in Scotland. The golden eagle is a bird of the hills in Scotland, and the authors have extensive knowledge of arctic-alpine ecology and land use, as well as personal acquaintance with most of the country's eagle home ranges and habitat. It is with this comprehension that they conclude by interpreting their findings in these papers and assess how golden eagles live in the human landscape now and might subsist in future.

NOTE:

Most of the data used in these papers was recorded in the authors' own times, except the information for north-west Sutherland, which was recorded by SR while working on studies of golden eagles with Jeff Watson. Some of the data, such as grouse and ptarmigan numbers, were noted by the authors when in eagle country for other purposes, such as other studies, or simply hillwalking or skiing. The authors are great believers in noting down as much data as possible whenever in the field, regardless of the reason, but passing and noticing another.

Accumulated information on the location and use of golden eagle eyries took many years to gather. In brief, the only way to ensure that all of them had been found in any area was to search every suitable crag and tree. Hence, the exact methods for nest finding and recording breeding success are not given here, although generally, they were as subsequently adopted and described in the national golden eagle surveys (Dennis *et al.* 1984, Hardey *et al.* 2006).

Commonly in animal behaviour, a home range is where an individual animal lives, and a territory is a smaller area where an individual dominates all others of the same sex. Golden eagles in each home range typically use different nests in consecutive years, on our study areas (Watson 1957, Brown & Watson 1964, Watson *et al.* 1989, Watson *et al.* 1992) and generally (Watson 1997). The nests usually occur in a cluster within the central part of the home range, one nest being used in each cluster each year. The distance between nearest neighbour nests in different clusters significantly exceeds that within clusters, for instance in our north-east area (Watson & Rothery 1986). Adults in a pair display aggressively to other pairs, which led Watson (1997) to use the word "territory". However, whether they occupy territories with the above definition is unclear, so we use "home range", and take the occupancy of each cluster as equivalent to the occupancy of a home range. Our criterion for deciding whether two pairs occupied two closely adjacent clusters rather than one pair taking over both of them is that a nest in each cluster must have held eggs or young at the same time. This conservative criterion is the only one that is reliable in the absence of marked birds.

References

Brown, L. H. (1969). Status and breeding success of golden eagles in north-west Sutherland in1967. British Birds 6, 345–363.

Brown, L.H. and Watson, A. (1964). The golden eagle in relation to its food supply. Ibis 106, 78–100.

Dennis, R.H., Ellis, P.M., Broad, R.A. and Langslow, D.R. (1984). The status of the golden eagle in Britain. British Birds 77, 592–607.

Hardey, J., Crick, H.Q.P., Wernham, C.V., Riley, H.T., Etheridge, B. and Thompson, D.B.A. (2006). Raptors: a field guide to survey and monitoring. The Stationery Office, Edinburgh.

RESAS (2015). Economic report on Scottish agriculture 2015 edition. Scottish Government, Edinburgh.

SNH (Scottish Natural Heritage) (1998). Annual Report 1997-98. SNH, Battleby, UK.

SNH (Scottish Natural Heritage) (2001). National assessments–recreation and access. SNH, Battleby, UK.

Watson, A. (1957). The breeding success of golden eagles in the north-east Highlands. Scottish Naturalist 69, 153–169.

Watson, A. and Rothery, P. (1986). Regularity in spacing of golden eagle Aquila chrysaetos nests used within years in northeast Scotland. Ibis 128, 406–408.

Watson, A., Payne, A.G. and Rae, R. (1989). Golden eagles Aquila chrysaetos: land use and food in northeast Scotland. Ibis 131, 336–348.

Watson, J. (1997). The golden eagle. Poyser, London.

Watson, J., Langslow, D.R. and Rae, S.R. (1987). The impact of land-use changes on golden eagles in the Scottish Highlands. CSD Report no. 720. Nature Conservancy Council, Peterborough.

Watson, J., Rae, S.R. and Stillman, R. (1992). Nesting density and breeding success of golden eagles (Aquila chrysaetos) in relation to food supply in Scotland. Journal of Animal Ecology 61, 543–550.

Contributors

ADAM WATSON was brought up at Turriff in Aberdeenshire. At age seven he became interested in snow and a year later in summer snow-patches in the Cairngorms, the largest expanse of high hill country in Scotland. Then at age nine his life changed after reading Seton Gordon's (SG) classic book *The Cairngorm Hills of Scotland*. Adam wrote to Seton soon after; he replied, and they corresponded for the rest of Seton's life. Adam was obsessively keen on the Cairngorms. At 13, alone, he saw his first pair of ptarmigan and thought they were the most beautiful birds in the world, more than eagles, his first pair of which he saw later that day. In 1944 he spent a day in the hills with SG, who showed him two eagle eyries, and Adam has been studying eagles in Deeside since. In the 1950s he met other eagle pioneers Charlie Palmar and Pat Sandeman. Leslie Brown especially made a big influence on what should and could be studied on eagles. Although Adam worked as a biologist, all his studies of eagles have been on his own time.

(photo Iain Cameron)

STUART RAE grew up in Aberdeen surrounded by coast, farmland, woods and moors, all within walking or bicycling distance. Exploring the outdoors and all wildlife was captivating to him, although birds were the most readily watched and so became the focal attraction. Gradually, he ventured into the hills; the pinewoods, long glens, cliffs and high plateau. Those hills, the Cairngorms, became and still are his favourite stomping ground. Many of those early hill days were spent with his older brother, Robert 'Skitts', and it was he who showed him his first eagle eyrie. There were two full-grown chicks perched on the edge of a huge eyrie in a Scots pine, an indelibly impressive sight to a 14-year-old. Stuart's teenage years were spent rock- and ice-climbing and studying birds throughout the Highlands, all the time gathering experience on eagles in different parts of the country. Then in 1982 he landed a dream job studying golden eagles with Jeff Watson, and he has been studying them professionally and privately since.

(photo Adam Ritchie)

The Cairngorms are an impressive hill range and eagles are impressive birds. They lured Adam and Stuart, eventually they met one another there. Early meetings were brief in passing and a bit one-sided, as Stuart was a schoolboy and Adam had, by then, an international reputation as an ecologist. Stuart's clearest reminiscence is of when he had topped out of a climb, appropriately named Eagle Ridge, on Lochnagar, a satellite peak of the main Cairngorms massif. Adam was there with climber and naturalist Tom Weir, having just watched him doing the climb, and all three had a chat about the magnificence of the route. In the 1970s, Adam had been enthused by Skitts ,who had started to study eagles, and they with Stuart began sharing information on them, a forerunner to the formation of the North East Raptor Study Group. In the late 1980s Stuart worked on the Montane Plateau Ecology Project run by Scottish Natural Heritage, where he studied dotterel, ptarmigan, snow bunting and other birds in the alpine zone. During that project Stuart began a PhD study of ptarmigan with Des Thompson, Ian Patterson and Adam as supervisors. Adam had meantime been studying corn buntings with Skitts, and later he continued the study far more intensively with Stuart.

Although each has his own experiences, they are continually pleased to learn from one another. Adam and Stuart have studied a lot together and have enjoyed each other's company on many fine days for the hill. Their discussions on various topics related to the hills and birds are strengthened by a shared familiarity with details, to such an extent that one can describe a place, a particular rock even, and the other knows where it is. They discuss methods and fieldwork intensively, relying on one another's individual perspective to complement and improve his own. Each has learned much from the other.

DEREK SPENCER became interested in eagles after retiring and moving to the north-west Highlands. As an experienced rock climber he was asked by members of the Highland Raptor Study Group to help gain access to some cliff eyries. He was immediately enthralled and inspired to continue more eagle study of his own. That in turn led to a demanding schedule, monitoring eagle behaviour in more than 20 home ranges in north-west Sutherland since 2012. Most of those home ranges are in the old study area used by AW in the 1950–60's and SR in the 1980s, and Derek has meticulously followed SR's methods. This has enabled comparisons of the feeding and breeding behaviour of golden eagles between the study periods as described in this book. Derek does most of this on his own in remote hills, at his own time and expense. SR first met Derek on a day hike over Foinaven, a magnificent hill in Sutherland. During their walk they covered a great tract of land checking for eagles, and their conversation covered a wealth of mutual and collective knowledge of eagles in the area. His inquiring mind, fitness and observance qualify Derek as a competent eagle observer. His honesty and humour make him a wonderful companion.

Acknowledgments

Adam Watson and Stuart Rae thank all fellow members of the North East Scotland Raptor Study Group (NESRSG), from the founders to those currently active, for recording and sharing information on eagles specifically, and for establishing and maintaining the strong database on all raptors in their study area. In particular we thank Skitts for showing Stuart his first eagles and then exploring eagle country, and for subsequently learning and sharing a wealth of knowledge on eagles. Skitts has coordinated the monitoring of golden eagles in the north-east since 1982 and has helped the authors tremendously with information over the years and for this book. Derek Spencer and Bernard Hendy shared their data, which helped to establish many of the results. Duncan Rae, Des Thompson and David Ellis improved the text with comments and corrections. We thank the former Nature Conservancy Council for funding research in the 1980s on golden eagles in the Scottish Highlands by the late Jeff Watson, and Derek Ratcliffe and Derek Langslow for supporting this work. All photographs are by either AW or SR unless otherwise stated, and we especially acknowledge AW's father AW senior and David Jenkins for their images.

EAGLE PEOPLE

Three authors who stand out in the history of the study of golden eagles in Scotland are Seton Gordon, Leslie Brown and Jeff Watson. They all graduated in natural sciences at university, zoology in the latter cases. However, they were of different times, and how they followed up their professional qualifications and studies of eagles reflected their times. Together, their work spanned a century and they were three of the main contributors to knowledge of golden eagles.

Seton Gordon (1886-1977), a Highland naturalist. A man of a time, who could be truly referred to as a pioneer in the study of golden eagles. Most of his writing was in popular, non-scientific books and magazines. He described eagles in the first chapter of his first book *Birds of Loch and Mountain* (1907) and he went on to write *Days with the Golden Eagle* (1927) and *The Golden Eagle* (1955). Adam Watson and Stuart Rae were inspired by his books to follow his path, explore the hills and search for eagles. He was, like AW and SR, born in north-east Scotland and, like them, his interest developed from encountering eagles during explorations of the local hills. Seton wrote 27 books on Highland wildlife, landscape and people, his breadth of knowledge revealing all life in the Highlands is integrated.

Leslie Brown (1917–1980), a raptor enthusiast. By profession a government agriculturalist who was more than an amateur ornithologist in his spare time. He spent most of his working life in Kenya, where he studied a variety of birds including the local species of eagles. But when he visited Scotland, his ancestral home, his attention was fixed on golden eagles. He instigated AW to join him in the first methodical study of the breeding density of golden eagles in Scotland. Their ground-breaking paper 'The golden eagle in relation to its food supply' (1964) described how the numbers of eagles differed with food available in regions with different environments across the Highlands. That work combined old-fashioned field explorations to find all the eyries with the scientific method of statistical analyses to test their findings.

(photo Peter Steyn)

Jeff Watson (1952–2007), a professional ornithologist. The son of an amateur ornithologist, Jeff studied zoology at the University of Aberdeen and went on to work for the Nature Conservancy Council (NCC), later Scottish National Heritage (SNH), in ornithological research and conservation. His father, Donald the wildlife artist, was an authority on hen harriers, and Jeff's PhD study was on the Seychelles kestrel. In 1982–85 he took on an ambitious study of golden eagles in the Highlands funded by the NCC. That study closely followed and was similar to an updated and expanded version of the Brown & Watson study, testing their earlier findings in nine ecological regions with different human land-use. SR was Jeff's assistant on that project, and between them they covered most of the Highlands. In 1997 Jeff wrote *The Golden Eagle* monograph, which was based on Scottish studies although it amalgamated information from throughout the golden eagle's Holarctic range.

(photo Vanessa Watson)

Two people who helped in our eagle studies died during the compilation of this book and are greatly missed. Eagles play an important role in our environment and society, and the world needs compassionate people who care for and understand the value of eagles. The world needs more people like Jenny and Bernard.

Jenny MSR Watson, wife of AW, who died on 7 September 2016 after 61 years of marriage.

Jenny helps look for golden eagle and ptarmigan on Derry Cairngorm in April 1955.

Bernard Hendy, friend of SR, who died of a heart attack in April 2016, while on a wildlife tour in Zambia.

Bernard in the Sutherland hills in 2014, where he monitored the golden eagles and other birds, in the 1960s and '70s.

(photo Simon Cherriman)

Looking down on a seven-week old eagle chick in its cliff eyrie. The prey on the edge of the nest are two mountain leverets and three red grouse. Sprigs of Scots pine, birch and aspen freshen up the nest.

1. NORTH-EAST SCOTTISH GOLDEN EAGLES AND LAND USE BEFORE AND AFTER 1937

Adam Watson & Stuart Rae

SUMMARY

This paper describes one local population of Scottish golden eagles *Aquila chrysaetos* in terms that can be considered for other local populations in a global context. The golden eagle has a wide range across the northern hemisphere, living in a variety of environments, mostly in open hilly or mountain areas. Its main requirements are medium-sized prey, such as rabbits *Oryctolagus cuniculus*, mountain hares *Lepus timidus* and red grouse *Lagopus lagopus scoticus*, and cliffs or trees to nest in. Although Scotland is a small country, it holds a diversity of habitats and food available for eagles. These affect the food eaten, the eagles' population density, and their breeding success. Human land use has had a world-wide effect on golden eagles, and in Scotland, particularly in the north-east, there has been considerable historical documentation of the eagle and how it has lived in changing environments. This paper documents such changes and effects on eagles in a long-term study area for eagles (e.g. Watson 1957, Watson *et al.* 1989, Watson *et al.* 2012). The year 1937 is used as a reference point because that was when a change in numbers of breeding eagle pairs was first recorded. Use of such information can help interpret how the golden eagle will probably fare in north-east Scotland in the future, and elucidate how eagles might be affected by changes in land use elsewhere.

EAGLES AND LAND USE UP TO 1937

Most of the open moorland that forms the bulk of the north-east Highlands lies below the potential timber line (about 600 m in the study area) and most of it was formerly forest. Before the forests disappeared, golden eagles would have been scarcer, as for example, on open Swedish mountains, where their density is about 2.5 times that on woodland-dominated land (Tjernberg 1985). Scottish deforestation led to moorland, which has been kept open ever since by burning and grazing. These moorlands form good hunting grounds for eagles. The heather *Calluna vulgaris* that has thus been artificially fostered provides abundant food for browsers and stays mostly green in winter, unlike most hill grasses, which die back. Heather abundance led to artificially increased numbers of red grouse and mountain hares. Carcases of deer, sheep and cattle on the open hill would have been more visible and accessible to eagles than in woods.

On a mild November day, a golden eagle gains height as it soars above the lower hills of the east Cairngorms—the heart of eagle country in north-east Scotland. The dark heather of the nearer hill is moorland, the paler hill behind being alpine ground used by ptarmigan.

Alluvial and freely drained soils on grassy glen bottoms and lower hillsides up to 450 m were formerly arable, supporting many farmers (Nethersole-Thompson & Watson 1981). Big patches of fine smooth grassland up to 750 m held many people in summer shielings (rudimentary dwellings), with cattle and sheep. People still used shielings until the late 1800s. No farmers have lived in some glens for many decades since, varying from the late 1700s to the mid 1800s, and in most glens since the late 1800s. When farmers ploughed the best valley soils and enclosed them with wooden fences, the red deer *Cervus elaphus,* which now depend on grasslands there for winter grazing, would have been much scarcer, because of the far smaller area of sheltered high-quality grazing available, and because farmers would have killed many deer (Nethersole-Thompson & Watson 1981, Watson 1983).

Deer carrion would have been scarcer too. Rabbits, which have abounded and been virtually unmolested by people on these grasslands since the 1940s, would have been scarcer on fenced arable fields, and farmers would have killed many there. In

The south side of the Cairngorms massif. Golden eagles occur for about 70 km from here to the edge of lowland farmland, 30 km from the city of Aberdeen on the east coast of Scotland. The hills are mostly rounded in shape, with many corries and alpine plateaux on the higher peaks and heather-dominated moorland on the lower hills. Such ground holds ptarmigan, red grouse and mountain hares, the three most commonly eaten prey species of golden eagles in north-east Scotland.

the late 1700s, the semi-open Old Caledonian woodland relics in upper Deeside today were young woods (Watson 1983), where hunting would have been more difficult for eagles, and capercaillie *Tetrao urogallus* were probably absent there temporarily, or extirpated. Moreover, the Earl of Fife wrote that large parts of what became the current study area were grassy in the late 1700s, with little heather, and red grouse scarce because of too much burning and poaching (Earl of Fife, unpublished diary 1783–92, available in Aberdeen University Library).

The mid and late 1700s would therefore have offered less food for eagles. Since then, eagle food has been more abundant, following human depopulation, management for red grouse and red deer, and the opening of the Old Caledonian woods by felling, natural tree deaths, and the lack of regeneration due to many deer. The fostering of red deer and their consequent increase in the late 1700s and early 1800s (Watson 1983) would, along with the under-shooting that has been general (Mitchell *et al.* 1977), have led to more deer carrion. However, eagles were killed on deer land during that time, e.g. at Mar Lodge. During his stay at Mar Lodge between 27 May and 9 June 1791, by which time deer had greatly increased from very low numbers in 1785, the Earl of Fife 'by way of exercise and amusement (sic) destroyed Eagles, Hawks etc.' (Watson 1983). These dates were in the middle of the breeding season, when eagles have young in the nests, and such killing would have cut breeding success. If this were an annual occurrence, the local eagle numbers were probably kept low. Given that the Earl was doing this, it seems likely that his 'foresters' (deerstalkers), who lived there all year round, also killed eagles.

In the 10 years 1776–86, 70 eagles were recorded killed in the two parishes of upper Deeside: Crathie and Braemar; and Glenmuick, Tullich and Glengairn (Old Statistical Account). Killing continued into the early 1800s. During 1831–34, 171 adults and 53 young and eggs were destroyed (New Statistical Account). "About this time and even later, the eagle was persecuted relentlessly, and it is indeed remarkable that the species should have survived" (Gordon 1955). People were paid bounties for eagles killed, but such bounty schemes in the 1900s were well known to be abused (e.g. Brown 1976), and so the totals in the vermin lists were inflated and unreliable. Nonetheless, there is no doubt that persecution of golden eagles in Deeside was severe in the late 1700s and early to mid-1800s.

In 1768, James Robertson wrote that eagles had become scarce in Skye because of persecution, but still infested north-west Sutherland (Henderson & Dickson 1994). At Durness, he noted in his journal of 1767 (Henderson & Dickson 1994) that the grass on the outer half of the peninsula of Farout Head was 'inclosed, for the purpose of receiving My Lord Rae's (i.e. Reay) newly weaned lambs, which are here safest from the attacks of the Foxes & Eagles that infest the country'. Of Skye he wrote in 1768, 'Not many years ago in this Isle they were much infested with Eagles and Foxes, but the Gentlemen agreed among themselves to raise a Sum of money yearly, which was bestowed in rewarding those who were most active in destroying

High eagle country in the Cairngorms, where snow lies on the hills well into spring. The upper limit of the tree line for tall trees is at about 550 m and the tops of the hills rise to 1309 m at the summit of Ben Macdui. The hill shown here is Beinn Mheadhoin (1182 m).

those, to every person who brought the scalp of a fox *Vulpes vulpes*, four shillings, for the head & feet of an Eagle, two & six pence, by those means they have rooted out those troublesome neighbours.'

By contrast, in the next century the Rev Thomas Grierson (1851) found that a Duke at Mar Lodge was protecting the eagle, presumably the Duke of Leeds. On a visit to Glen Derry, he wrote, 'The eagle does not always build among rocks. On a solitary pine, I was shown a nest in which young ones have been reared for many years, for, unlike most sportsmen, the Duke protects these noble birds, being unwilling they should be extirpated, of which there seems much danger.... Eagles have not been detected destroying the young deer, though they often fly after and annoy them, apparently in sport'.

Clutches of golden eagle eggs in the Royal Scottish Museum were taken on 'Lochavon Rock, Ben MacDhui', in April 1876 and 1877. A pair nested on crags at the upper end of the loch up to the late 1910s (IG, SG, DM, see below). No nest has been seen there since. This is near a 1910s range north of Cairn Gorm and a range to the south, occupied during 1942–1964 (Watson *et al.* 2012). Grimble (1896) wrote of the deer-forest of Glenavon that it held 'some of the highest hills in Scotland, and on their steep sides several pairs of eagles nest each season'.

Pioneering naturalist William MacGillivray made some revealing remarks about the rarity of golden eagles. He died in 1851, his book being published by order of Queen Victoria after his death (1855). Hence his observations would have been mostly before 1850. He saw an eagle soaring over the side of Ben Avon during a botanical trip to Creag an Dail Bheag (page 130), and in reference to a visit to Glen Callater with a local guide, he wrote 'Few birds occurred in the course of our excursion (page 140). Red Grouse, our host informed us, are much less numerous than they were twenty or thirty years ago. A few Eagles and Ravens are occasionally seen.

Much of the heather moorland in the eastern Highlands is a human artefact, perpetuated and exaggerated, especially on moorland managed for driven-grouse shooting. On grouse moors, this is usually done in small patches at a time, creating a mosaic over many years. The old tall heather provides cover for roosting and nesting grouse, while the short recently burned areas hold short sprigs of fresh nutritious growth for the grouse to eat.

There are many Plovers on the hills at the head of the glen, and sometimes Dotterels are met with there'. On page 391 he gave a detailed account, as follows: 'Golden Eagle. This species, which formerly existed in considerable numbers in Braemar (his reference to Braemar here and elsewhere means the parish), and bred in the precipices of the wilder glens, is now very seldom to be seen there. Shepherds and gamekeepers have effected its almost entire destruction, insomuch that it is doubtful if even a single breeding-place remains occupied. In the course of six weeks' excursions among the mountains I saw only two individuals, one at Craigandal (Creag an Dail, Ordnance Survey OS), the other in Gleney—after all, it might have been only one individual twice seen.

Mr. Cuming has a preserved specimen, which was obtained upwards of twenty years ago. Mr. Brown, at Micras, who has others, has frequently had individuals sent him to be preserved. Mr. McGregor informed me that it is occasionally, though rarely, seen in Glen Callater; and Mr. Stewart has sometimes seen it over and about Ben Aun, and in the upper part of Glen Gairn. Several persons have mentioned its occurrence in the tract between Castletown (i.e. the eastern part of Braemar village) and Glen Tanar; but farther down the country it is scarcely ever to be seen. My friend, Mr. Thomas Jamieson, who has favoured me with ornithological observations made by him in Braemar, in July, 1846, mentions having seen an Eagle near Balmoral. The gamekeeper at Beallach-bhui Cottage (Garbh Allt Shiel, OS), informed him that eighteen years ago there was an Eagle's nest near the Garvalt (Garbh Allt, OS), on a tree, but that it was now forsaken. He also stated that a great many years ago, Eagles had bred on a neighbouring hill called Craig-an-dain (Craig Doin, OS). Braemar is thus evidently not the place for studying the habits of Eagles.'

Red grouse flushed during a stalk alarm stags with their calls and whirring flight and sometimes induce them to run away. This makes the red grouse unpopular with stalkers and may have been one of the reasons for the protection afforded to eagles by the Duke of Leeds (above) around 1850. Interviews with old stalkers revealed that a later tradition under the unusually enthusiastic deer-shooter the Duke of Fife (owner of the Forest of Mar in 1879–1912) was that red grouse be treated as a nuisance. In the early 1900s, he ordered that staff coming across nests of red grouse should smash the eggs and protect eagles because they killed red grouse. Informants have told AW that most staff did not destroy grouse nests in the Duke's time and none did so from 1930 onwards.

If one imagines a situation without human persecution in upper Deeside, numbers of golden eagles would have been low when there was probably a small amount of deer carrion prior to the mid 1780s. Then the numbers of eagles would probably have increased in response to more carrion food in the 1790s, 1800s and early 1900s. However, severe persecution would have prevented this, and it is obvious from MacGillivray's remarks that the eagle was at a very low ebb there in the mid 1800s. This probably continued until the Duke of Fife ordered protection of eagles during his ownership from 1879 until 1912. The eagles would then have increased, and protection occurred generally on the deer land in the early 1900s and thereafter until the 2000s. Hence the late 1940s and 1950s were probably very favourable to eagles there, because of abundant food in association with traditional deer management, combined with few competitors.

Some glens in north-east Scotland hold old woodlands, dominated by Scots pines with a mixture of birch, rowan and juniper. Eagles hunt over these woods and dip through openings to pounce on prey. The trees offer secure nest sites for them, and they are seldom far from open moorland or high hills where they hunt mountain hares and grouse. Closely-planted trees in dense plantations offer no hunting ground, and the authors have seen few eyries in such trees, only in very old plantations adjacent to moorland.

A typical Cairngorms glen; a steep-sided glaciated valley with a flat floor and a meandering river. There are no currently used eagle eyries in this particular scene, although eagles do nest on crags and trees in such landscapes. The eagles hunt ptarmigan on the high rocky terrain and mountain hares in the same high ground or in the lower heather-clad slopes. Red grouse live in most heather-covered ground apart from that on the tops and ridges, where it grows too short for the grouse to hide under. Red deer carrion and calves are more commonly found on the low ground. Rabbits live in some such glens, although mostly farther downstream in more sheltered glen bottoms, often in grassland where there are remains of abandoned human settlement and agriculture.

EAGLES AND LAND USE SINCE 1937

Watson *et al.* (1989) reported changes in the number of adult eagle pairs over the years during 1944–80, with an increase in the late 1940s followed by a decline in the 1960s. Each eagle pair used several alternative nest-sites in different years. These formed a cluster of nest-sites, with a relatively regular spacing between pairs, such that the mean nearest-neighbour distance of nests between clusters significantly exceeded that within clusters (Watson & Rothery 1986). The colonisation by new pairs was in a sequence, the subsequent abandonment being the reverse sequence, so that the last home range to be colonised was the first to be abandoned (Watson *et al.* 2012).

There was a base level of three pairs in the early years, with two new pairs colonising and building nest clusters nos. 4 and 5 (Watson *et al.* 2012) in the lower parts of the area in 1937–39. This followed an unusually high and prolonged abundance of red grouse during the mid 1930s in the Cairngorms region and Scotland generally (Mackenzie 1952, Hudson 1992). Also in the mid 1930s, rock ptarmigan *Lagopus muta* abounded in the Cairngorms massif (much of which is in our study area) in the mid 1930s (Watson 1965, Watson *et al.* 1998). After these peaks, red grouse and ptarmigan declined in the late 1930s. Ptarmigan increased to a new peak in 1940, with fairly high numbers still in 1941 and 1942, and red grouse were also common on the study area in 1939–42 (above references, AW unpublished, and informants JB, CG, AM, DM). In 1942 a new eagle pair no. 6 colonised. Red grouse and ptarmigan then declined to an unusually low trough in 1944 (Watson 1965, Watson *et al.* 1989).

Despite the very low densities of red grouse and ptarmigan in 1943–45, the three extra eagle pairs stayed, all on the lowest parts of the study area. However, rabbits were abundant there at all seasons every year, during 1943 (AW unpublished), and in 1944–45 as recorded by Watson *et al.* (1989). Also, a new factor arose there, following clear-felling of mature plantations and some old woodland during the 1939–45 war. A large Canadian lumber camp, built between eagle nest-clusters 4 and 5, operated until 1945, resulting in about 5.4 sq km of pine-wood being felled, which probably enlarged the hunting area for the local eagles. The ranging be-haviour of golden eagles can be greatly increased by their use of a felled area, as compared with before felling (Walker *et al.* 2005).

The wartime fellings for the Canadian lumber camp resulted in a large increase of open ground (Watson 2002). For five decades afterwards, that ground remained open because red deer ate tree seedlings, and their heavy grazing prevented heather and grass from becoming tall and dense. These conditions induced increased numbers of red grouse, mountain hares and rabbits (and in other areas since, AW unpublished), and would have afforded much easier access to hunting eagles than the woodland prior to felling. The numbers of red grouse would have probably increased too, as in a recent study in Argyll, where the numbers of red grouse increased on ground that had been clear-felled of coniferous plantations (Walker *et al.* 2005). We suggest that the increased prey on the felled areas sustained the extra two eagle pairs at clusters 4 and 5. This would have compensated for the trough densities of red grouse and ptar-migan on moorland and alpine land at greater distances from the nests. A sixth pair colonised ground away from the felled areas in 1942, but later colonisation by anoth-

Young pine trees are slow-ly becoming established in Glen Derry, Deeside (May 2016), subsequent to large reductions in the number of red deer in the area since 1995. The authors have in-cluded in this volume pho-tographs of scenes re-taken after a gap of more than 50 years, to illustrate changes in eagle country over the course of time. It would be useful if this scene were taken again in 50 or more years to see whether this change continues.

er pair (no. 7) in 1947 was also in the lower parts of the area and close to nest-clusters 4 and 5. This settlement is attributed partly to the same reasons.

The nests built by pair 7 were also near the same open expanses created by wartime felling. Furthermore, by 1947 the numbers of red grouse and ptarmigan were increasing on moorland and alpine land (Watson et al. 1989). After the wartime felling, groups of red deer that previously wintered in sheltered woodland then had to live on open ground with poor shelter. Heavier mortality would be expected in those groups, and thus more carrion. This would have been additional to the counts of dead deer (Watson 1971, Watson et al. 1989), which were not carried out on the felled areas. Although the felled land comprised a tiny proportion of the study area, for red deer it had formed a large proportion of the most sheltered wintering grounds at low altitudes. Also it formed a high proportion of the land near nest-clusters 4, 5 and 7.

Watson (1957) described a concentration of three pairs with cases of occupied nests only 2.4 km apart in the same year, and in one year three occupied nests spaced as a triangle with sides 4, 2.4 and 2.8 km long. We now report that this involved clusters 4, 5 and 7, and the triangle was of nests on opposite sides of two valleys. Of the five extra new pairs that colonised, therefore, three of them built nests in clusters in a small proportion of the study area, but this formed a large proportion of the lowest ground. It had the greatest densities of rabbits, as well as the increase of red grouse, mountain hares and rabbits on the sections that had been felled.

The new pair 8 colonised a higher part of the study area in 1948. This followed a large increase of red grouse and ptarmigan after a very low trough (Watson et al. 1989, Watson et al. 1998). Deer carrion was exceptionally abundant in the snowy winters of 1947, 1951 and 1955, but quite frequent in other years (above references and Watson 1971). The number of deer shot annually was increased during the war to help offset the national shortage of meat for human consumption, e.g. attaining the high total of 275 stags shot in 1940 (Whitehead 1960). In the post-war late 1940s and 1950s, by contrast, far fewer were shot, averaging only 175 stags per year (Watson 1983), only 64% of the 1940 total. This low level of shooting would have led to heavier compensatory mortality by starvation, resulting in an unprecedented number of dead deer, according to the stalkers. These years of more food on high ground in the late 1940s would have provided an opportunity for the extra pair of eagles to colonise and breed.

The first abandonment site, cluster 8, was the last to be colonised. It followed a decline of deer carrion in the late 1950s, which was also associated with several mild winters. Also, the numbers of ptarmigan and red grouse declined in the late 1950s to a low trough, and very few rabbits occurred in the upper glens following localised extinction after the introduction of myxomatosis in 1955 (Watson 1957, Brown & Watson 1964, Nethersole-Thompson & Watson 1974, Watson et al. 1989). By 1961, densities of ptarmigan and red grouse had risen again, but although rab-

The highest ground in the north-east Scotland study area; the summits of Cairn Toul (1291 m) and Sgor an Lochain Uaine (1258 m) seen from the summit of Braeriach (1296 m), May 2013. Eagles hunt these hill tops and their flanks, although this high ground is often obscured by cloud, and the frequency of birds hunting there is probably under-recorded. The eagles usually hunt here by gliding along the hillsides, purposely ambushing ptarmigan or hares as they round ridges or bluffs. In winter, when snow cover can be almost 100%, the ptarmigan form large flocks, often with hundreds of birds, and they flush to fly off when an eagle swoops towards them. They must present a challenge for the eagles to catch. This is also true of red grouse and moorland, though the near 100% snow cover is less common there.

bits soon increased after the localised extinction, they remained scarcer than in years before myxomatosis arrived. The amount of deer carrion decreased greatly from 1964 onwards (Watson 1971), following a large increase in the proportion of the deer population that was shot annually, and more supplementary feed given to deer in winter and spring.

These effects were exacerbated by applying fertilisers on hill grassland to increase grass production for deer, and by allowing the deer access to grass fields on fertile arable soils. The abandonment of clusters 6–8 is probably attributable to reductions of carrion, an important winter food for eagles on the study area (Watson 1957, Brown & Watson 1964, Watson *et al.* 1989, Watson *et al.* 1992).

Subsequently, a pair colonised cluster 6 again in 1971. This coincided with the greatest abundance of ptarmigan, red grouse and mountain hares recorded in the study area since the mid 1950s (Watson *et al.* 1989, 1998, 2000). In addition there came localised increases of prey animals due to tree planting on areas spanned by clusters 2 and 4–7. Seventeen coniferous plantations were established on heather moorland, starting in the late 1960s and continuing into the 1970s, with ploughing to increase the eventual nutrient uptake, and fertiliser applications at the time of planting. This led to a short-lived boom in the numbers of red grouse, black grouse and particularly mountain hares inside the deer-fenced plantations, until the canopy closed. Hares numbered scores in each plantation and hundreds in the largest one (AW unpublished, and local deerstalkers RS and WF), to the extent that stalkers shot scores of them in a day. Increases of red and black grouse have been documented in other cases of tree planting on moorland in northern Scotland (Parr 1992, Baines *et al.* 2000), and increases of mountain hares have been widespread in recently-planted coniferous plantations on moorland in northern Scotland generally (AW unpublished).

The abandonment of cluster 6 after its re-colonisation in 1971 followed a very deep decline of ptarmigan and red grouse, and lower numbers of mountain hares (Watson *et al.* 1989, 1998, 2000), as well as extirpation of red grouse and mountain hares due to canopy closure and extinction of under-storey heath by dense shading. Canopy closure also reduced accessibility for eagles to hunt. This suggests that low densities of the main prey species cannot sustain a high density of eagle pairs, a conclusion already found from detailed studies in several Scottish regions (Watson *et al.* 1992). They examined average differences between areas, not differences within an area between years, as in the present study. However, as shown here, the fact that the same conclusion came from both kinds of comparison makes it the more likely explanation.

By the late 1980s and early 1990s, the increasing density of red deer and their annual heavy grazing had converted much heather to short smooth grass and made grassy vegetation even shorter (Watson 1989, 1990). Both factors favoured rabbit grazing. This in turn led to a further spread of short grass and short heather, so that the heavy grazing by deer and rabbits combined to increase the nutritive value of the pasture for both species, as well as making the pasture ever shorter and thus more favourable to rabbit foraging. Formerly, rabbits had been restricted to grassland on alluvial glen bottoms and grassy patches on nearby lower slopes. Now they spread on to heathery slopes that formerly had held none because the vegetation was too tall and too lacking in grass.

In addition, muirburn fires on the most heavily grazed ground induced a fresh sward based on grass rather than heather (Watson 1990). This was adverse to red grouse and mountain hares, whose main food plant is heather. It favoured an expansion of the distribution of rabbits as well as a rise in their abundance.

As a consequence of these changes, the original estimate of 25 sq km suitable habitat for rabbits (Brown & Watson 1964) increased after 1985 to an estimated 44 in the late 1980s and early 1990s. However, deer-fencing for dense tree planting destroyed good grazing due to dense shade, and deer-fencing for natural tree regeneration made other habitat unsuitable for rabbits because ground vegetation grew tall and heather replaced grass. The reduction of rabbit habitat by plantations and fencing for natural tree regeneration amounted to 14 sq km, hence a drop from 44 to 30. In addition, heavy culling of red deer in the 'regeneration zone' of Mar Lodge Estate (where this plot is situated) reduced deer densities in that zone, and in turn this induced an increase of tall heather, a replacement of grass by heather on grass-heather swards, and on grass swards on alluvial soils an increase in the height of the grass and heath. All these are inimical to rabbits. The loss of good rabbit hab-itat is estimated at 9 sq km due to tall heather and 3 sq km due to increase in grass height, so an overall fall from 44 to 18, or a drop from the original 25 to 18 sq km. Not only have these changes led to rabbits becoming more scarce, but the increase of tall heather is also adverse to red grouse and mountain hares.

LIKELY FUTURE CHANGES IN EAGLE NUMBERS

Current management of red deer in those north-central parts of the study area where natural tree regeneration is the main aim involves all deer seen being shot. The superabundance of deer carrion seen in 1947 and the early 1950s is therefore very unlikely to recur, and the high density of breeding pairs of eagles in those years on the north-central parts of the area is also unlikely to be seen again.

The area of cleared woodland has increased since 1997, when felling commenced of the alien coniferous plantations established in the late 1960s and 1970s. More than counteracting this, however, much larger deer-fenced areas have allowed natural tree regeneration since 1993 on lower ground, leading to increases of tall heather and thickets of young trees, combined with no muirburn and no cutting of heather, thus leading to low densities of red grouse, mountain hares, and rabbits. Although these changes would be expected to lead to more black grouse, these are not a useful substitute for red grouse and mountain hares and rabbits, because the proportion of black grouse in food items at and near eagle nests in the area is so much lower than the proportions for each of the other three species individually and much smaller than for the other three species combined (Watson–Chapter 4). The evidence for the number of lekking blackcocks on Mar Lodge Estate, which in-cludes the study area, indicates that many blackcocks were counted at leks in spring 2005 and 2007, but with no conspicuous sustained increase (Francis & Pout 2005, Littlewood 2007). Also, dense tree-planting on open moorland has resulted in re-duced eagle numbers in other Scottish regions (Marquiss *et al.* 1985, Watson 1997).

In the future, therefore, any large increase of woodland and its associated land use of no muirburn and no cutting of heather would be expected to reduce the num-ber of eagle pairs breeding in the study area. Because cluster 2 lies in the midst of an area earmarked for large-scale woodland expansion, with cluster 4 almost as central, it seems likely that these two clusters would most likely be abandoned, and in that order. At cluster 1, only a small proportion of low-altitude hunting ground has been lost due to natural tree regeneration so far, but there is a growing threat of abandonment because of cessation of muirburn and promotion of reforestation. If these increases in tree cover were to develop, this would likely leave only one pair (cluster 3) on the study area instead of the original three. Pair 3 would persist large-ly, but not entirely, on high treeless areas. Even cluster 3 seems less secure now that natural woodland is being strongly promoted (e.g. Cairngorms Partnership 1999).

The likelihood is that one or two pairs would continue, albeit with larger home ranges because there would still be some hunting ground left open for the eagles. Although it is likely that this would be in more isolated patches, with fragmented populations of the eagles' main prey, grouse and hares, on open moorland and smaller numbers of alternative prey in the wooded areas. This is in line with an assessment of possible effects of native woodland expansion on the birds of the Cairngorms by Whitfield & Tharme (1997), who stated 'Only red grouse and, to a lesser extent, golden eagles should be negatively affected to a marked degree'.

In 1997, under the European Union's Birds Directive, part of the Cairngorms area, which includes this study area, was designated as the Cairngorms Special Protection Area (SPA), within which the golden eagle is one of the qualifying species. Then in 2010, a specific area, overlapping the previously mentioned SPA and also including the study area, was designated as the Cairngorms Massif SPA, with the golden eagle being the sole qualifying interest. Under the Habitats Directive there is also a Special Area of Conservation, within which the native pine-wood is one of the qualifying habitats. Although any future expansion of native pine-wood would increase the botanical value of the SAC, conflicts of interest and of legal obligations would be inevitable if that expansion were to occur at the expense of golden eagles.

Golden eagles are well known to be susceptible to disturbance, so these designations might be seen as a welcome approach to their protection. Inadvertent disturbance has long occurred on the study area, and in some cases led to immediate desertion by incubating eagles, or to eggs becoming too cold and subsequently failing to hatch under incubating birds (Watson 1957, Watson *et al.* 1989). Nests near roads and tracks in the Cairngorms region have tended to fall into disuse as tourist numbers rose (Watson 1981), and this tendency has become more obvious in recent decades, along with an increase of tourism. Entirely new conspicuous paths have been made to encourage access by people. One route which has been mapped to encourage visitors to walk in a previously seldom visited area passes within 200m of one eagle nest. Although the development in that case was done in ignorance of the eagle nest's proximity, that eyrie has not been used since the 1980s. New paths and conspicuous upgradings of existing paths now proliferate on several hills hunted by eagles, and if the number of people using them continues to rise, this would probably have a detrimental effect on the eagles' hunting behaviour. A new potential threat is that the Cairngorms National Park Authority, established in 2004, is putting much effort into increasing tourism across the park, including the study area (CNPA 2006). This poses grave threats to the remaining eagles, through increased human disturbance at and near the nests.

To conclude

We have documented the circumstances of golden eagle population changes in a part of north-east Scotland, following an increase and then decline in their numbers over more than a century. The decreasing amounts of deer carrion and prey; and the increases of woodland and human disturbance all pose threats to the remaining pairs of golden eagles breeding on the study area. Although the historical high density of breeding golden eagles in the study area is interesting, it is distinctly unnatural, associated with prehistoric massive deforestation and lesser deforestation during the 1939–45 war, and aided shortly afterwards by large amounts of deer carrion. Subsequent changes in deer management led to less deer carrion and a fall in eagle numbers. The remaining pairs will probably face threats from large-scale reforestation and human intrusion in this complex area of international importance for nature conservation. If climax forest and scrub ecosystems are attained, there would still be golden eagles breeding, though no longer at high density.

References

Anonymous (2004). Raptor round up 2002. Scottish Birds 24 Supplement.

Baines, D., Blake, K. and Carradine, J. (2000). Reversing the decline: a review of some black grouse conservation projects in the United Kingdom. 2000. Cahiers d'Ethologie 20, 217–234.

Brown, L. (1976). British birds of prey. Collins, London.

Brown, L.H. and Watson, A. (1964). The golden eagle in relation to its food supply. Ibis 106, 78–100.

Cairngorms National Park Authority (2006). Cairngorms National Park Action Plan. CNPA, Grantown-on-Spey, Moray.

Cairngorms Partnership (1999). Cairngorms forest and woodland framework. CP, Grantown-on-Spey, Moray.

Francis, I. and Pout, A. (2005). Black grouse in north east Scotland in 2005. North East Scotland Bird Report 2005, 116–119.

Gordon, S. (1955). The golden eagle. Collins, London.

Grierson, T. (1851). Autumnal rambles among the Scottish mountains. James Hogg, Edinburgh.

Grimble, A. (1896). The deer forests of Scotland. Kegan Paul, Trench & Trübner, London.

Hudson, P.J. (1992). Grouse in space and time. The Game Conservancy Ltd., Fordingbridge, Hampshire.

Henderson, D.M. and Dickson, J.H. (Eds) (1994). A naturalist in the Highlands: James Robertson, his life and travels in Scotland 1767–1771. Scottish Academic Press, Edinburgh.

Littlewood, N. (Ed) (2007). North East Scotland Bird Report 2007, 28.

MacGillivray, W. (1855). The natural history of Deeside and Braemar. Printed for private circulation, Bradbury & Evans, Printers Extraordinary to the Queen, London.

Mackenzie, J.M.D. (1952). Fluctuations in the numbers of British tetraonids. Journal of Animal Ecology 21, 128–153.

Marquiss, M., Ratcliffe, D.A. and Roxburgh, R. (1985). The numbers, breeding success and diet of golden eagles in southern Scotland in relation to changes in land use. Biological Conservation 34, 121–140.

Mitchell, B., Staines, B.W. and Welch. D. (1974). *Ecology of red deer: a research review relevant to their management in Scotland. Institute of Terrestrial Ecology.*

Nethersole-Thompson, D. & Watson, A. (1974). *The Cairngorms: their natural history and scenery.* Collins, London.

Nethersole-Thompson, D. & Watson, A. (1981). *The Cairngorms: their natural history and scenery.* Melven Press, Perth.

Parr, R.A. (1992). *Moorland birds and their predators in relation to afforestation.* PhD thesis, University of Aberdeen.

Tjernberg, M. (1983). Prey abundance and reproductive success of the golden eagle, *Aquila chrysaetos*, in Sweden. *Holarctic Ecology* 6, 17–23.

Walker, D., McGrady, M., McCluskie, A., Madders, M. and McLeod D.R.A. (2005). Resident golden eagle ranging behaviour before and after construction of a windfarm in Argyll. *Scottish Birds* 25, 24–40.

Watson, A. (1957). The breeding success of golden eagles in the north-east Highlands. *Scottish Naturalist* 69, 153–169.

Watson, A. (1965). A population study of ptarmigan (*Lagopus mutus*) in Scotland. *Journal of Animal Ecology* 34, 135–172.

Watson, A. (1971). Climate and the antler-shedding and performance of red deer in north-east Scotland. *Journal of Applied Ecology* 8, 53–67.

Watson, A. (1981). *Detailed analysis. Evidence at the Lurcher's Gully Public Inquiry.* Nature Conservancy Council, Aviemore.

Watson, A. (1983). Eighteenth century deer numbers and pine regeneration near Braemar, Scotland. *Biological Conservation* 25, 289–305.

Watson, A (1989). Recent influences on grouse habitat from increases in red deer. *Annual Report* 5: 45-46. Joseph Nickerson Reconciliation Project, Thornhill.

Watson, A (1990). Land use, reduction of heather, and natural tree regeneration on open upland. *Annual Report, Institute of Terrestrial Ecology*, 25–27.

Watson, A. (2002). The main changes in the Cairngorms since 1938. *Mountain Views* 50, 17–21.

Watson, A. and Rothery, P. (1986). Regularity in spacing of golden eagle *Aquila chrysaetos* nests used within years in northeast Scotland. *Ibis* 128, 406–408.

Watson, A., Moss, R. and Rae, S. (1998). Population dynamics of Scottish rock ptarmigan cycles. *Ecology* 79, 1174–1192.

Watson, A., Moss, R. and Rothery, P. (2000). Weather and synchrony in 10-year population cycles of rock ptarmigan and red grouse in Scotland. *Ecology* 81, 2126–2136.

Watson, A., Payne, A.G. and Rae, R. (1989). Golden eagles *Aquila chrysaetos*: land use and food in northeast Scotland. *Ibis* 131, 336–348.

Watson, A, Rae, S. and Payne, S. (2012). Mirrored sequences of colonisation and abandonment by pairs of golden eagles *Aquila chrysaetos*. *Ornis Fennica* 89, 229–232.

Watson, J. (1997). *The golden eagle.* Poyser, London.

Watson, J., Rae, S.R. and Stillman, R. (1992). Nesting density and breeding success of golden eagles (*Aquila chrysaetos*) in relation to food supply in Scotland. *Journal of Animal Ecology* 61, 543–550.

Whitehead, G.K. (1960). *The deer stalking grounds of Great Britain and Ireland.* Hollis & Carter, London.

Whitfield, D.P. and Tharme, A.P, (1997). *An evaluation of the effects of native pinewood regeneration on the birds of the Cairngorms.* Unpublished report, Scottish Natural Heritage, Edinburgh.

NAMES AND AREA OF WORK OF ESTATE STAFF WHO GAVE INFORMATION TO AW

Robert Brown, Balmoral

Donald Campbell, Invercauld

Eck Duff, Glen Shee

Robert Esson, Dinnet

James Gillan, Invercauld & Balmoral

Willie Gillanders, Glen Gairn & Dinnet

George Gordon, Abergeldie & Balmoral

Sandy Grant, Glen Shee

Alexander J. Grant, Corndavon

Charles Grant, Mar

Ian A. Grant, Mar

William Grant, Mar

Dave Gruer, Glen Shee

Sandy Gruer, Glen Shee

Ronnie Hepburn, Invercauld, Glen Shee

Alexander J. McDonald, Mar & Invercauld

Donald McDonald, Mar & Invercauld

William L. McGregor, Balmoral

Donald McHardy, Abergeldie & Balmoral

Colin F. McIntosh, Invercauld

James McIntosh, Corndavon

Peter Mackintosh, Mar

Ian A. McLaren, Mar

Thomas McPherson, Invercauld

Charles Milne, Glen Tanar

John Robertson, senior, Glen Muick & Bachnagairn

John B. Robertson, Abergeldie, Glen Muick & Balmoral

Willie Ross, Dinnet & Glen Gairn

Frank Scott, Mar

Robert L. Scott, Mar

Ronald Scott, Mar

Walter Scott, Mar

Charles A. Wright, Abergeldie & Balmoral

Jock Wright, Invercauld

2. NEST SITES OF GOLDEN EAGLES IN NORTH-EAST SCOTLAND

Adam Watson & Stuart Rae

SUMMARY

Each home range of a pair of golden eagles *Aquila chrysaetos* on the study area held more than one nest, varying from four to 12, with a mean of 9.4. Most were in crags, some in trees, and a few on the ground on steep slopes. A high proportion of crag-nests faced from north to east, as did the crags available. Most of the area lacked woodland, and trees on cliffs were too small to hold nests. Many crags stood at higher altitudes than trees that were big enough to hold nests, and eagles built many crag-nests at altitudes far above the uppermost big trees. However, they avoided crags above 900 m, where much snow lay in spring. During years with big snowstorms in April–June, birds reared more young per undisturbed pair in tree-nests than on crags.

INTRODUCTION

The aims of this paper were to report the numbers of nests and features of nest-sites used by pairs of golden eagles on a study area in north-east Scotland during 1943–80. The eagles were known to have a varied number of nests per home range; some home ranges contained nests in trees but most did not. It is typical in golden eagles internationally (Watson 1997) and on the study area (Watson 1957) for each pair to use different nests in different years. Nests in any one home range tended to form a cluster, with significantly shorter distances to the nearest neighbouring nests within clusters than between clusters (Watson & Rothery 1986). There was also known to be much variation in the aspect of nests on crags and in the slopes where eagles had nests in trees. Crag is used in this context because the word describes not only steep or high cliffs, but small rocky outcrops which eagles often use as nest sites.

The above papers, together with Watson (1957) and Watson *et al.* (1989) described the area and methods for finding the numbers of pairs and nests on it. Eagle eyries can remain in recognisable form for many years, often for decades, after use and all these were counted during the study. The work was mainly in upper Deeside, but with much effort in the Angus glens and some in upper Banffshire and north-east Perthshire. Most pairs and nests were found in the early years by AW. Data on numbers found in successive years gave some confidence that a near total enumeration was probably achieved. The aspects of nest crags and the slopes where nest trees

A large golden eagle eyrie fills the crown of an ancient Scots pine. The nest is about 2m in diameter, more than a metre deep and over 20 m up the tree. The tree is probably at least three hundred years old and still in good health. Eagles have used this nest since the 1980s. The first eyrie that AW knew was not far from this site, in the same eagle nesting territory, and that nest had been used since the end of the 19th century, as known by Seton Gordon.

were growing were assessed to the nearest 45°, and all orientation statistical analyses of nest sites were done in Orion 4 (Kovach 2012).

The nest heights and the heights of the nest trees were not measured, because Scots pine *Pinus sylvestris* is the main species of tree used by nesting golden eagles in the north-east (Watson 1957) and there is much variation in the growth forms and heights of Scots pines in the study area, such as different soil types and exposure to wind (Steven and Carlisle 1959). Eagles used trees about 25 m tall in sheltered valley woods and trees of less than 10 m in thin soil on hillsides near the upper limit of the tree growth. Many eyries were in such places, where exposure plus isolation has led to short gnarled trees with strong branches.

RESULTS

NUMBER OF PAIRS

Although AW checked some pairs in 1943 and 1944, the study began properly in 1945, when he found 22 adult pairs on land managed for deer-shooting or deer and grouse-shooting. No more were found in 1946 and 1947, but two new immature pairs appeared in 1947–48 and started breeding during 1949–50 in sites that had held no nests in 1945–48.

NUMBER OF NESTS

There were 313 nests on crags, including three on very steep slopes, and 86 in trees, of which there were 85 and 37 respectively in the intensive study area. Most (83, 96%) trees used for nesting were Scots pine, and one each of Norway spruce *Picea abies*, European larch *Larix decidua* and downy birch *Betula pubescens*.

Across 11 ranges that were occupied each year in 1945–47 (1–11 in Table 1 of Watson 1957), 59 nests were found in 1943–45. Later, AW learned that two nests which he discovered in 1946 and 1947 had already been known to

An eagle eyrie on a sheltered but precarious branch, which has since fallen to the ground. Eagle eyries in trees are usually supported by more than one branch, which helps to support the increasing weight of nests that are added to each year and become very heavy.

Young Jenny Watson (AW's daughter, July 1968) stands below the eagle eyrie illustrated in Seton Gordon's book The Golden Eagle, King of Birds, *perhaps the most renowned golden eagle eyrie in Scotland (the world?).*

the deerstalkers in 1943–45.

On the area of intensive study, five ranges 1–5, 32 nests were found in 1945–47, an old one in 1948, and an old one in 1951 when the first complete check of all trees and crags was done. This revealed some nests that the deerstalkers had not seen, but no pairs or nest-clusters that were unknown to the stalkers. During each year in 1951–57, all crags and trees were searched except in 1953. All possible eyrie sites were again searched in 1964. These total searches revealed no extra nests. However, one new nest was found in 1973 and another during 1976, in sites that had not held nests in earlier decades.

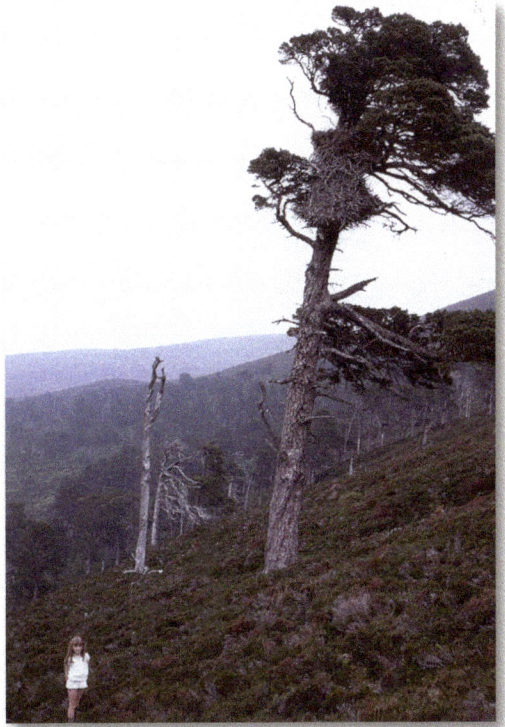

In 1945–47, in the extensive area with its eight pairs, 27 nests were found and then for each successive year 3, 0, 2, 2, 0, 0, 0, 0, 0 additional nests, totalling 34. In subsequent years there were 2, 2, 0, 0, 0, 1, 4, totalling 43, 4 of which were new nests as judged by their construction with all-fresh material. Therefore, 79% of the total number of nests must have been present in the first three years (as known from later observations up to 1950), compared with 97% in the intensively searched area.

Almost all nests at the start of the study had been built in previous years, and eagles did not use some of them for egg-laying in subsequent years. At a few of these, no signs were found in subsequent years of them even having been touched up with new sticks. However, most nests listed in Table 2.1 were used, all of them in ranges 1, 2–4, 7–8, 11–15, and 19. During the study, vegetation over-grew some nests on moist north-facing ledges, and grassy patches remaining there are now the sole sign of where the nests once stood. Gales blew down some nests in Scots pines, and snow avalanches tore out a few on crags. Others on sunny freely-drained ledges and in trees are still there, although several have not been used since 1950, and two not in any year during the study.

Comparison of a golden eagle nest tree in Deeside over 57 years: Left, June 1959. (David Jenkins) Right, June 2016. (SR)

This eyrie is in a Scots pine at the upper limit of the altitude where such old pines grow, as remnants of once more widespread and vigorous woodland. Limbs have fallen off, including most of the nest branch, which happened in the late 1970s. The boulders in the early shot had been exposed by muirburn when the local deer stalker, Bob Scott at the time, burned the tall rank heather to promote fresh, more nutritious growth for the red deer to eat. Such burnings for deer were often large and whole hillsides could be burned at once, unlike the smaller burnings used for grouse moor management. The heather in the later shot is taller since the old burning practice has been stopped, and the number of deer has been reduced. There is no sign of regeneration of seedling trees in the photograph, although there are young pines beginning to grow above the height of the heather lower on the same hillside. This is another photograph which the authors would like to see re-taken in 50 years or more.

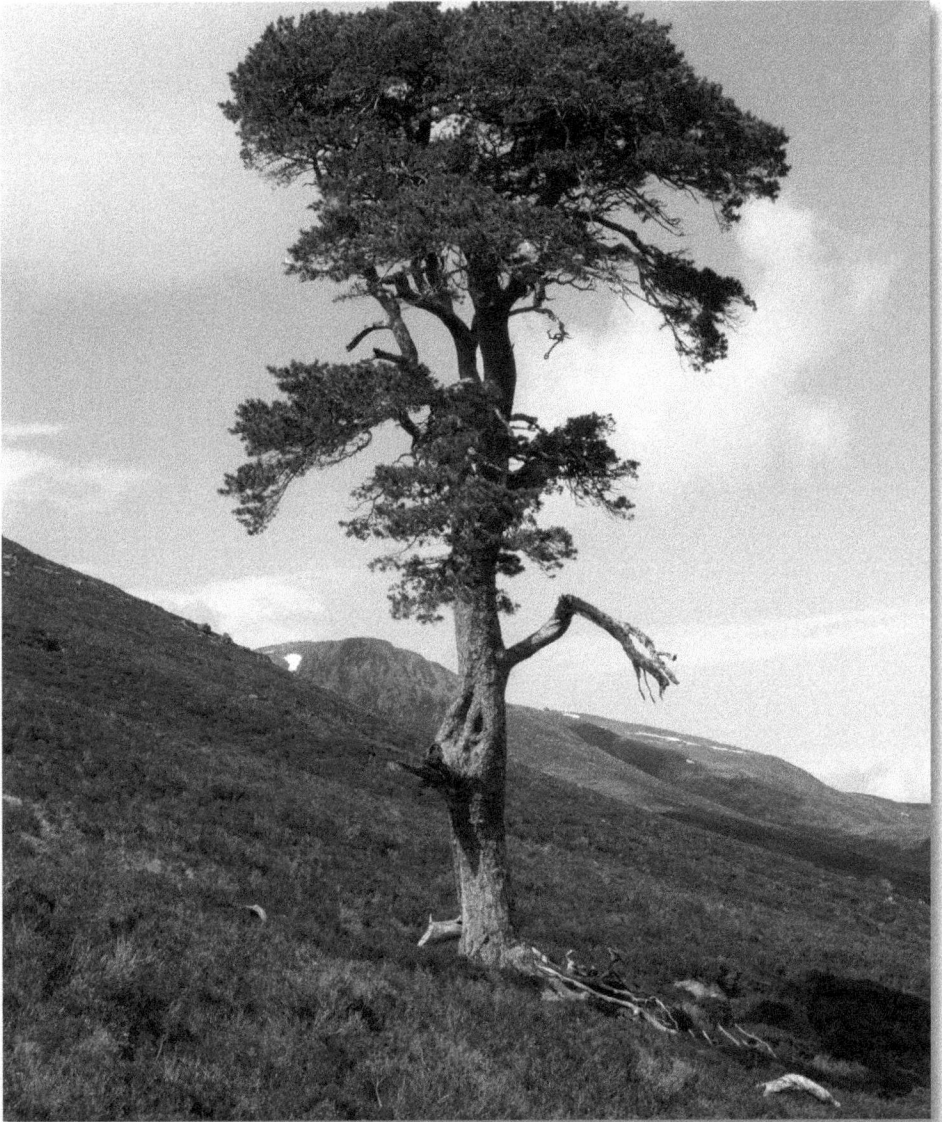

In clusters 1–5, studied in all years, the largest number of nests used in a cluster was 12 (Table 2.1). In no cluster did eagles lay in the same nest for more than two years running, except in cluster 3, where birds used a tree-nest in four consecutive years. In the main study area, all 13 nest clusters had crag nests and only three did not include tree nests (Table 2.1). In the 53 clusters of the wider study area, seven did not contain crag nests and 34 did not have tree nests (Table 2.2).

NEST BUILDING AND CONSTRUCTION

An eagle was seen carrying a green pine branch to a nest in late October, and adults often added sticks in November–December. Thereafter, building increased in frequency each month to March. In many instances, birds added a few sticks to one

Comparison of a golden eagle nest tree in Deeside over 69 years:
Left, Capt. George F. Raeburn and AW stand below an eyrie, 20 July 1947. (AW senior)
Right, Ewan Weston and Skitts stand below the same nest tree, 01 May 2016. (SR)

Sometime since 1947, the nest has been relocated to the branch above the former one. Note the tall heather in second image, which is a consequence of the removal of high numbers of red deer in the area since the mid 1990s. The subjects in the former picture stand only ankle-deep in heather, while those in 2016 stand with it up past their knees. Such is the effect of heavy grazing by high numbers of herbivores. The new thick stand of trees in the background in the latter year is a result of a tree plantation in the 1970s. The perimeter fence that had been placed around the plantation to protect the growing trees from deer has been removed, hence the effect of a wall of dense woodland emerging abruptly from the heather moorland. The people in the second shot could not stand in quite the place as the former couple, as there was a large branch fallen from the nest tree filling the exact spot. George Raeburn was a solicitor in Ellon, Aberdeenshire, and was still in the army after the war. As a keen ornithologist, he assisted AW in the early days of his eagle studies and later helped with AW and V.C. Wynne-Edwards to form the North-East Branch of the Scottish Ornithologists' Club, the SOC's first branch.

alternative nest (in two cases to two alternative nests) before the breeding season, including in a few instances as early as December. Birds that apparently did not lay eggs often put sticks on several nests, sometimes building a nest completely, and in spring added branches with fresh green shoots. In one pair where the hen was immature, as assessed by white areas in her flight feathers, birds placed two sticks on a nest in one year and none in another, with only a few pieces of lining in both years.

All nests in Scots pines consisted of large pine branches. Many occupied nests on crags were made of pine branches, but in every case some pines grew alongside.

The scale of an eagle eyrie in a tall Scots pine is illustrated by the addition of human figures for comparison. Here, Ewan Weston passes down an eagle chick to be ringed and measured safely on the ground by his wife Jenny. This eyrie is on Mar Lodge Estate in Deeside, which is run by the National Trust for Scotland, and Shaila Rao, ecologist with the NTS, looks on.

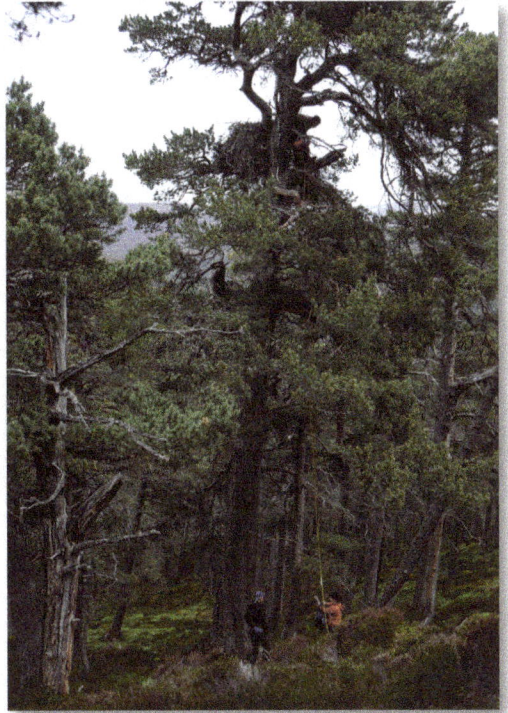

On crags more than 200 m from trees, nests were composed mainly of heather stems. Other species of tree from which eagles often added a few sticks were juniper *Juniperus*, larch and birch, and in a few cases the nearest such trees grew at least 3 km away.

Nests in pines had linings mostly of green pine shoots, and occasionally in addition a few larch shoots or leaves of great wood-rush *Luzula sylvatica*, while some had fresh shoots of larch, birch, rowan *Sorbus aucuparia*, and green sprays of heather *Calluna vulgaris* or sometimes of blaeberry *Vaccinium myrtillus*. One nest with young in 1946 contained a piece of rotting plank 45 cm long, 8 cm wide and 2 cm thick, perhaps taken from a dilapidated wooden hut in the glen bottom below the nest.

Eagles often took fresh branches with green leaves of larch or birch to the nest during the incubation period and occasionally later, till at least early July. They usually removed these green sprays when leaves began to wilt, and then brought another fresh spray. At one nest this happened frequently in April and May, even though the hen was immature and did not lay eggs.

Most nests were about 1 m deep and a few up to 1.5 m, but two consisted of a few sticks loose on a ledge, with eggs placed on growing grass in the centre. One nest in a pine was 6 m deep, stretching from less than half way up the tree to the very top. In the last year when the birds used the top part, before a gale blew out most of a section underneath, the eaglets' heads rose above the topmost shoots. After the gale, the eagle still used this site, but laid eggs in a new cup 3 m from the foot of the pile, with the uppermost part forming a partial roof that afforded much shelter from rain.

USE OF NESTS
Within a given nest-cluster, eagles used some nests more than others. There was no

Figure 2.1. Aspects of nests on crags and trees in home ranges 1–13

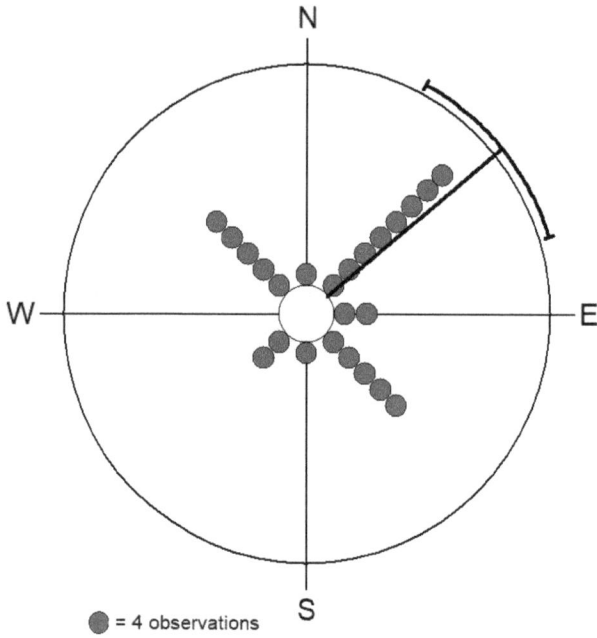

Figure 2.1(a). Aspects of crags where nests were on rocks (n = 85) mean vector = 50°, SE = 11.4°, with 95% confidence interval.

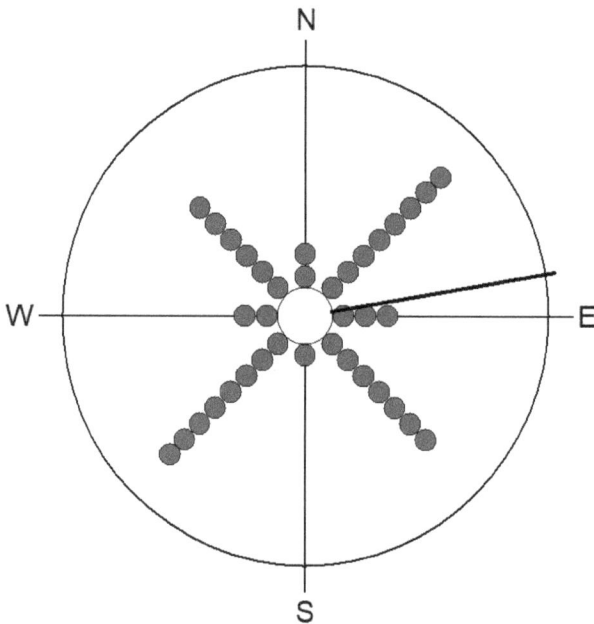

Figure 2.1(b). Aspects of hillsides where nests were in trees (n = 37) mean vector 80°, SE = 138.6°.

Figure 2.2. Aspects of nests on crags and trees in the extensive study area.

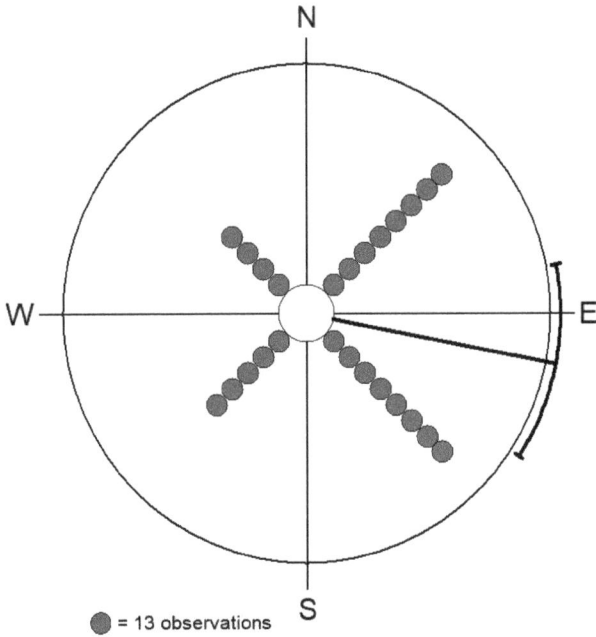

= 13 observations

Figure 2.2(a). Aspects of crags where nests were on rocks (n = 313) mean vector = 97°, SE = 12.4°, with 95% confidence interval.

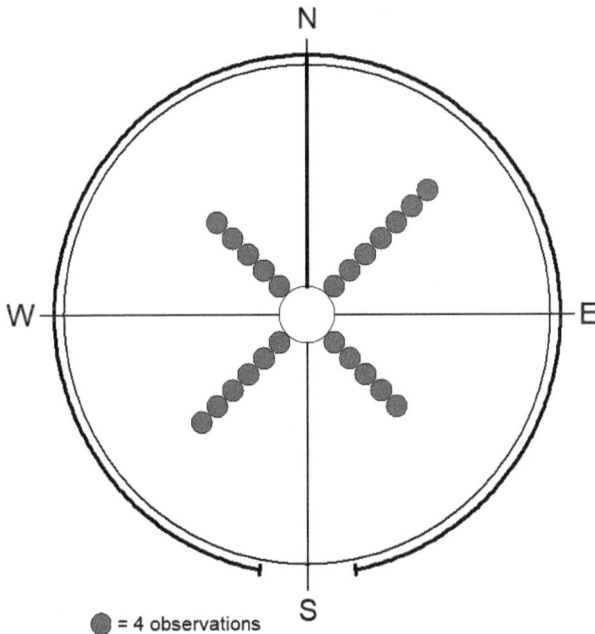

= 4 observations

Figure 2.1(b). Aspects of hillsides where nests were in trees (n = 86) mean vector = 0°, SE = 86.2°. with 95% confidence interval.

obvious reason for this. Overhanging rocks completely sheltered some nests from rain, but others stood on exposed rocks and trees.

Nests used every year might possibly become foul, but eagles in cluster 3 reared five young despite the same hen using the same tree-nest four years running. Eagles whose eggs were often robbed, such as in clusters 4 and 11, had no more nests than in clusters 2 and 10 where they were never robbed during the study.

One explanation for the use of many alternative sites is that this part of Scotland receives heavy snow that would bury some nests in spring. However, the evidence went against this. The number of nests per nest-cluster with all nests on crags above 600 m (2, 6, 7 and 8), did not surpass that per cluster at lower altitudes. Apart from cluster 2, on open moorland, the four clusters with the most nests (4, 5, 9 and 10) had some nests in pines, which were less affected by deep drifted snow than crag ledges. Moreover, in years of exceptional snowfall, such as spring 1951, eagles in cluster 4 bred on a crag at 880 m, where spindrift snow blown off a nearby plateau piled up on all the crag ledges, yet the eagles had three available nests in pines at 520 m with far shallower snow on the ground and little or none in the trees.

ALTITUDE AND ASPECT OF CRAG AND TREE NEST-SITES

On the area's high treeless land, all the nests in five clusters (2, 6, 7, 8, and 11) were on crags, whereas Scots pines held about half the nests in the other seven clusters. Even where nests lay on crag ledges, birds often built immediately behind a small tree, which supported the outer part of the nest. Such trees also afforded shelter from wind and precipitation, and shade from strong sunshine at some sites. The lowest nest in the intensive area, a tree eyrie, lay at 390 m altitude and the highest was on a crag facing south-east at about 900 m. Opposite to the latter and in the same nest cluster, a nest stood on a crag facing north-east at 870 m. In the extensively studied area, which covered lower land to the east and south-east, some nests were at much lower altitudes, and none stood at such high altitudes as the highest ones on the intensive area. The lowest nest on the extensive area was in a tree at 170 m.

Eagles had most crag-nests on rocks facing from north-west through east to south-east (Tables 2.3 & 2.4). The aspect of slopes where trees held nests showed a more even spread, though with some concentration on slopes facing south-west or north-east. In the core area, home ranges 1–13, most crag eyries were on rocks facing from north-west through east to south-east, with a mean aspect of 50°, significantly different from that expected in a random distribution (Rayleigh Test, Z = 11.74, P < 0.001, n = 85). Tree eyries were not placed on hillsides of any preferred aspect (Z = 0.08, P = 0.92, n = 37) and they had a mean aspect of 80° with a wide standard error of 139° (Figure. 2.1).

In the extensive study including home ranges 1–13, most crag eyries were on rocks facing from north-west through east to south-east, with a mean aspect of 97°, significantly different from that expected in a random distribution (Z = 10.50, P < 0.001, n = 313). Tree eyries were not placed on hillsides of any preferred aspect (Z = 0.22, P = 0.80, n = 86) and the hillsides had a mean aspect of 0° with a wide standard error of 86° (Figure 2.2).

SUCCESS FROM NESTS IN CRAGS AND TREES DURING SEASONS WITH SNOWSTORMS
During snowstorms with drifting, some nests on crag ledges were exposed to blown snow falling down the crag and piling up deeply on ledges, to the risk of avalanche, and to the actual snowfall itself. Being at lower altitudes than crag-nests, tree-nests tended to receive more rain than snow. Also, any snow on tree-nests melted more quickly, again owing to lower altitude and longer daily exposure to the sun.

In three years with severe April–June snowstorms, 1975, 1977 and 1979 (AW diaries unpublished), the numbers of young reared per undisturbed tree-nest were 1, 2 and 2 in 1975, 0, 1, and 3 in 1977, and 0, 2 and 2 in 1979. In contrast, undisturbed crag-nests in 1975 showed 9 failures and 3 cases of 1 reared; in 1977 there were 3 failures, 5 cases of 1 reared, and 3 cases of 2 reared. In 1979 there came 3 failures, 4 cases of 1 young reared, and 2 cases of 2 reared. When collated across years, the results indicated poorer success from crag-nests than tree-nests (Mann-Whitney U-test; U = 84, P = 0.036, n = 33,9). This might suggest the value of tree-nesting during a snowy spell. However, most tree-nests stood at lower altitudes than most crag-nests, and so were probably less adversely affected by snow for that reason alone.

Table 2.1. Number of nests (n = 122) on crags and trees at 13 ranges in 1943–80

Range*	On crags	On trees	No. seen unused since 1944**	No. seen unused, but estimated used 1939–43	Range occupied in all years
1	1	5	0	0	+
2	12	0	2	2	+
3	6	4	1	1	–
4	3	5	4	4	+
5	7	3	2	2	–
6	10	2	3	1	+
7	11	0	0	0	+
8	6	3	0	0	+
9	8	3	2	1	–
10	6	4	2	2	+

11	8	4	0	0	–
12	3	4	0	0	–
13	4	0	0	0	–
Total	85	37	15	12	

* Ranges 1–12 as in Watson (1957, Table 1).
**Built in years before the present study began.

Table 2.2. Number of nests of golden eagles on cliffs and in trees for 53 ranges outside the main study area. East Highlands. Ranges with very small numbers of nests such as 1 or 2 nests were occupied only ephemerally, on grouse-moors, and the birds usually soon vanished.

Range	On crags	In trees	Total
14	16	0	16
15	13	0	13
16	11	0	11
17	9	0	9
18	9	0	9
19	4	0	4
20	2	1	3
21	4	0	4
22	11	0	11
23	8	0	8
24	5	0	5
25	8	1	9
26	1	1	2
27	6	0	6
28	10	0	10
29	4	0	4
30	5	0	5
31	6	0	6
32	3	2	5
33	3	1	4
34	7	0	7
35	10	3	13
36	7	1	8
37	6	0	6
38	4	0	4
39	3	0	3

Table 2.2 Continued			
Range	On crags	In trees	Total
40	1	1	2
41	1	0	1
42	2	0	2
43	0	5	5
44	2	2	4
45	2	0	2
46	5	1	6
47	4	0	4
48	0	2	2
49	0	2	2
50	0	5	5
51	1	7	8
52	4	0	4
53	4	0	4
54	4	0	4
55	0	4	4
56	5	0	5
57	4	0	4
58	2	0	2
59	4	0	4
60	0	4	4
61	2	0	2
62	3	0	3
63	2	0	2
64	0	1	1
65	2	0	2
66	1	2	3

Table 2.3. Percentage of nests observed on cliffs of different aspect and in trees on slopes of different aspect, on ranges 1 to 13 as in Table 2.1

	n	N	NE	E	SE	S	SW	W	NW
Crags	85	4	36	8	23	1	8	0	22
Trees	37	5	22	8	19	3	22	5	16

Table 2.4. Percentage of nests on crags facing different aspects, and of nests in trees on slopes facing different aspects, combining data for the 13 ranges on the main study area and the 54 ranges on the extensive area

	n	NE	SE	SW	NW
On crags	313	30.6	32.9	19.8	19.5
In trees	86	30.2	19.8	26.7	23.3

DISCUSSION

The most golden eagle nests found in any cluster in the main north-east study area was 12 nests, but there were 16 in one cluster in the wider area and there might have been more because some sites may have been long unused and might not have been identified. Some home ranges in Scotland contain more nests, e.g. there are 17 nests on one cliff in one home range (Rae 2012). Old deerstalkers knew of five trees with nests in the 1940s which had also been used for nests before 1895. Indeed, most crags that the eagles used for nests, and many trees that they used for nests, are traditional, as are the ledges or branches with the nests. Even sites apparently used for the first time known to us may have been used some years earlier, before winds or avalanches blew or tore nests off, or plants on the nutrient-rich site overgrew the nests. At one nest last used in 1952 and not blown or torn out, no trace of a nest remained in 2002, and only a tiny mound of grass and moss marked the spot.

Watson (1957) gave data showing that most nests on crags in the study area were on crags facing from north-west through north to east, but tree nests had a more even spread of slope aspects. The concentration of aspects in the crag sites was to be expected, because most crags in the study area, as in the north-east Highlands and the Scottish Highlands generally, face these directions. This is because most crags resulted from the last glaciers having formed in hollows that faced these directions (Bremner 1912, Barrow *et al.* 1912, Sissons 1972). Writing about a frequent claim that most nests face north, Brown (1976) noted correctly that the preponderance of crags with a northerly aspect resulted from glaciation.

Finding that nests face north disproportionately, Watson & Dennis (1992) compared the proportions of slope aspects at nests with the proportions of slopes of different aspect taken randomly from maps. They concluded that eagles chose slopes whose aspects differed significantly from those expected from the proportions available, and tended to prefer north-facing slopes and avoid southerly aspects. This ignored Brown's explanation. Later, Watson (1997) reported that Cliff Henty had sent him a letter stating that the analysis by Watson & Dennis was invalid because it should have used crag aspects on maps, not slope aspects. Also, Henty noted that a high proportion of crags in Scotland faced aspects from north to north-east, due to past glaciers. Watson (1997) admitted that Henty's argument 'has

Although eagle eyries can be large, not all are so. This is an example of a small nest built on a cliff in the Cairngorms. The nest is barely noticeable from below, and from the topside, it can be seen to be not much more than a ring of heather placed at the back of a ledge. The eyrie is high above the tree line, so the most convenient material for nest building is heather. This is less substantial and does not knit so well into a solid framework as tree branches, especially those of Scots pine. Therefore, the nest is more readily blown apart by strong mountain winds. Also, the nest is not used every year, and less maintenance leads to more deterioration. The overhang above the nest gives some protection to the sitting birds and their young, but not enough to prevent the effect of weather on the nest when not in use. However, it is clear that the same ledge has been used by generations of eagles; the vegetation below the eyrie has been enriched by the nutrients seeping from the birds' droppings and food remains. The blaeberry on the nest ledge is vibrant green and a rich growth of mosses trails down the rock beneath.

merit', but in fact Henty and the studies by geomorphologists refuted the validity of the analysis in Watson & Dennis. The present study has confirmed that eagles in the north-east of Scotland nest on crags with north-to-southeasterly aspects, but when nesting in trees, the trees grew on no particular hillside aspect.

There was an advantage for eagles which nested in trees because they had a higher breeding success than those on crags in years of heavy snowfall. Tree nests were often at lower altitude than crag nests and were probably received more sunshine and hence warmer temperatures than nests on shaded crags. Any difference in breeding success at tree and crag nests in similar home ranges was probably related to the micro-climate at each nest site. However, we cannot discuss this any further, because rigorous comparisons of breeding success at different nest sites would require assessment of the micro-climate at each nest, which we did not do. Some home ranges did not contain crags and some did not have trees tall enough to nest in, but of those that held both, perhaps the eagles use tree-nests where suitable large trees are available, simply because tree sites are at lower altitude and hence get less snow and better shelter. Also, nests in pine-woods are more difficult to find than crag-nests, and this may be a traditional advantage for eagles on estates where gamekeepers have persecuted them.

Golden eagles probably first settle in an area where there is ample food. If there is enough food for them to breed, they will pair and then select a nest site. Those nests sites would then probably be the best available as assessed by the eagles for shelter from weather and security from potential predators, including man. Any conservation of eagles should encompass all their requirements and so should take these nesting criteria into account.

ACKNOWLEDGMENTS

We thank R. Rae, E. Duthie, J. Chapman, J. Hardey and R. Macmillan for some field notes.

REFERENCES

Barrow, G., Hinxman, L.W. and Craig, E.H.C. (1912). *The geology of the districts of Braemar, Ballater and Glen Clova. Memoirs of the Geological Survey of Scotland.* HMSO.

Bremner, A. (1912). *The physical geology of the Dee valley. Aberdeen University Press, Aberdeen.*

Brown, L.H. (1976). *British birds of prey. Collins, London.*

Kovach, W.L. (2012). *Oriana – circular statistics for Windows, ver. 4. Kovach Computing Services, Pentraeth, Wales, U.K.* http://www.kovcomp.com/oriana

Rae, S. (2012). *Eagle Days, Landford Press, Peterborough.*

Sissons, J.B. (1972). *The last glaciers in part of the south east Grampians. Scottish Geographical Magazine 88, 168–181.*

Steven, H.M. and Carlisle, A. (1959). *The native pinewoods of Scotland. Oliver & Boyd, Edinburgh & London.*

Watson, A. (1957). *The breeding success of golden eagles in the north-east Highlands. Scottish Naturalist 69, 153–169.*

Watson, A. and Rothery, P. (1986). *Regularity in spacing of golden eagle Aquila chrysaetos nests used within years in north-east Scotland. Ibis 128, 406–408.*

Watson, A., Payne, A.G. and Rae, R. (1989). *Golden eagles Aquila chrysaetos: land use and food in northeast Scotland. Ibis 131, 336–348.*

Watson, J. (1997). *The golden eagle. Poyser, London.*

Watson, J. and Dennis, R.H. (1992). *Nest site selection by golden eagles Aquila chrysaetos in Scotland. British Birds 85, 469–481.*

3. DEMOGRAPHY OF GOLDEN EAGLES IN SCOTTISH DEER-LANDS, 1943–1985

Adam Watson & Stuart Rae

SUMMARY

W e studied the demography of a golden eagle *Aquila chrysaetos* population in 1943–85 on an area of large private estates in upper Deeside where the main focus was shooting red deer *Cervus elaphus* and, to a lesser extent, shooting red grouse *Lagopus lagopus scoticus*. Eagle food was chiefly red grouse, ptarmigan *Lagopus muta*, mountain hare *Lepus timidus* and rabbit *Oryctolagus cuniculus*, and carrion from red deer. The annual number of adult pairs was correlated positively with the annual amount of carrion. Annual breeding success, defined as the number of young reared per undisturbed adult pair, was correlated positively with prey amounts in the same spring, especially with red grouse and rabbits. It was correlated negatively with the number of adult pairs. The annual proportion of adult pairs considered not to lay eggs was negatively correlated with the amount of prey, especially rabbits, and with breeding success in the previous year. The amount of dead deer in the spring was correlated negatively with the previous winter's air temperature and snow-lie.

INTRODUCTION

Most studies of golden eagles in Scotland have been short-term assessments of mean eagle density, breeding success, and food supply in different regions (Brown & Watson 1964, Watson *et al.* 1992).

Long-term fieldwork in upper Deeside (Watson 1957, Watson *et al.* 1989) differed by stressing between-year fluctuations in eagle density, breeding success and food supply within one area. Because amounts of prey and carrion vary greatly from year to year on any one area, long-term work helps achieve reliable con-

A golden eagle nest with two eggs on Mar, Deeside (April 1948), high up a glen in the heart of the north-east Scotland study area. In Scotland, golden eagles generally lay in late March, when snowfalls can still occur and cover any birds sitting on eggs on such high mountain crags.

clusions. Watson *et al.* (1989) recorded changes in eagle density across 37 years up to 1980 inclusive. Here we extend the data five more years to 1985 inclusive.

Watson *et al.* (1989) and Watson and Rae (Chapter 1, this volume) reported that an increase of breeding pairs in the 1940s was associated with increases of prey and of carrion from red deer *Cervus elaphus*. Then a decline of adult pairs in 1959–71 followed a large reduction in the amount of carrion of red deer, associated with a higher proportion of the deer population being shot. Shooting apparently compensated partly for previous large numbers of deer dying emaciated in spring, especially after severe winters when food for them was scarce due to snow.

It has long been well known in large raptors, including golden eagles, that some adult pairs do not lay eggs each year (Brown & Watson 1964, Watson *et al.* 1992). There has not been a satisfactory explanation for this, largely owing to studies on any one area being generally too short-term. Watson *et al.* (1989) found that the proportion not laying was associated negatively with the estimated weight of prey in spring, but not significantly (r = 0.293, P < 0.1). Below we address this question again, using a longer run of data. The null hypothesis is that the proportion of pairs not laying eggs does not differ between years of abundant food and little food.

Another question is whether breeding success falls with rising population density. This has been well established in research on many animal species, and, if present, it indicates a potential regulatory effect in limiting further rises in density. This has generally not been addressed in golden eagles, partly because they live so long and partly because most studies have been short-term. A recent study in the western Italian Alps showed breeding success falling as population density rose (Fasce *et al.* 2011). In their case, the number of pairs increased from one in 1972 (clearly close to complete extirpation and presumably due to human persecution) to 144

Looking down on a golden eagle eyrie, high in a Scots pine in Deeside. A clutch of three is rare in Scotland and the authors have only seen or heard of three-egg clutches laid by two other birds, which were also in the eastern Highlands, one on an adjacent home range to this nest. The hen that laid these three eggs was recognisable between years by her white tail markings and the colouring of her eggs. She first bred at this nest site in 1987 and so far, as of 2016, she had reared a total of 46 chicks in 29 years. The mean brood size was 1.6 chicks per annum, and in four years she reared three young. In 2016 she was still on territory, but no eggs were recorded.

Eaglets usually lie low and quiet when an observer approaches the nest, and smaller nests can be easily overlooked if the nest is placed in the tight crown of a pine, as is this nest. The far edge of the eyrie is firm and abrupt where it fills the gap in the canopy, and the near edge forms a long sloping ramp – the landing platform. Eagles fly low on approaching their eyries, swooping up and stalling in flight just as they touchdown on the landing ramp. Then they walk up onto the rim of the nest cup. Here the remains of a blackcock lie at the head of the ramp where a bird has delivered it recently and the hind quarters of a mountain leveret lie in the nest cup.

in 2008, a 144-fold increase in 36 years. With such a massive increase, it would be surprising had there been no relationship between breeding success and density. In our case in upper Deeside, the number of pairs did change between years, but on a minor scale and not due to human persecution, and hence likely to be more typical of established populations of this long-lived species. Thus our results are likely to be more typical across the world range. Our null hypothesis was that the number of young reared per adult pair would not differ between years in which the number of pairs differed.

STUDY AREA

Larger study areas in upper Deeside were described in some earlier papers (Watson 1957, Watson *et al.*1989, Watson *et al.*1992, Watson *et al.* 2012). The area for the present paper is smaller, the 360 sq km 'core area' of Watson *et al.* (1989). Our reasons for using it are that we knew it and its eagles better, and that measurements of prey amounts and deer carrion were made within it. It lies in uppermost Deeside, up to the highest summits of the Cairngorms range. It was all within one private estate until 1962 and was thereafter parts of two estates.

The chief economic interests of landowners comprised the shooting of red deer, and since 1962 a lesser interest in shooting red grouse *Lagopus lagopus scoticus*.

(Left) Bob Scott, a deer stalker on Mar estate at an eagle eyrie (June 1948). Bob lived with his family in a cottage far up what was then a remote glen, surrounded by eagle country. He knew all the eagle eyries and kept an interest in their well-being. In his youth, AW worked with Bob during the deer stalking season, and AW studied ptarmigan on the hills above Bob's home. They knew each other very well, sharing their knowledge of the hills and the local wildlife, including that of the eagles. (Right) AW picks up an eaglet to ring it, Mar (June 1948). Photograph taken by Bob Scott on same day as previous photograph.

Because of the area's high value for nature conservation and scenery, more than 80% of the estate is covered by national and international nature conservation designations. Most of it is moorland with much heather *Calluna vulgaris*. The moorland was forest in prehistoric times, during the warm period after the last glacial ice melted, but has since been kept open as moorland by burning and by deer and other browsing animals eating young trees. The valley (locally glen) bottoms have alluvial soils which supported some marginal arable cropping until farmers were evicted in the 1700s and early 1800s. The lower slopes in several glens carry stands of old Scots pine *Pinus sylvestris* and downy birch *Betula pubescens*, some coniferous plantations, and fenced enclosures for natural tree regeneration. Alpine land rises above 760 m altitude.

Descriptions of this area in more detail have been published elsewhere, along with data on the eagles' food supply (Watson 1957, Brown & Watson 1964, Watson *et al.* 1989, Watson *et al.* 1992), and information on home range, nest clusters involving alternative nests, and accounts of the attitudes of local deerstalking staff towards eagles (Watson & Rothery 1986, Watson *et al.* 2012). During the 42 years of the present study, local staff did not persecute eagles.

Two eagle chicks in a cliff eyrie in Mar (8 June 1947). This was the most successful of AW's pairs for breeding success. Both eaglets were reared. A dead fox cub lies on the eyrie, in the foreground. The whole prey-list that day consisted of a red grouse, a ptarmigan, a brown rat, and two young foxes each about a foot and a half long. This pair had a habit of killing foxes, much to Bob Scott's delight! This nest was set in a cave with a big overhang that kept the chicks dry in rain and snow. AW has fine memories of this nest and wrote a long account in his diary, including ringing the chicks. The black and white film is good for the contrast between the largely white birds and the dark rock. Taken with AW's father's box Kodak.

METHODS

Methods for finding the number of adult pairs and their nests, and for studying their breeding success, including the criteria for determining 'non-laying', the proportion of adult pairs considered not to have laid eggs, have been described fully elsewhere (references in above paragraph). Watson *et al.* (2012) added an important proviso, that nests were considered to be in use by separate breeding pairs only when adjacent nests held eggs or young at the same time. We use that proviso in the present paper. It is the only one that can be regarded as reliable in the absence of data on individually known birds, such as colour-ringing and study of DNA in feathers. A few breeding failures followed inadvertent disturbance near nests (e.g., logging and an artist painting) and egg-collecting (Watson 1957, Watson *et al.* 1989). Human-induced failures formed a small proportion of the total. Nevertheless, we looked for trends in 'number of young reared per undisturbed adult pair', as well as 'number of young reared per adult pair'.

AW made his first observations in 1943 and surveyed nest occupation by all pairs in the study area annually during and after 1945, using methods described elsewhere (Watson 1957, Brown & Watson 1964, Watson *et al.*1989). Watson *et al.* (1989) documented numbers and breeding success of golden eagles on the study area for 1944–80, and AW held the data that formed the basis of summaries for 1981 (Watson 1982). SR began observations in the mid 1970s and was employed to monitor numbers and breeding success on and around the study area in 1982–85 (Watson *et al.* 1992).

Methods for measuring the abundance of the eagles' four main prey species have been described in detail (Brown & Watson 1964, Watson & Hewson, 1973, Watson *et al.* 1973, Watson *et al.* 1989, Watson *et al.* 1998, Watson *et al.* 2000). The data comprised the number of adult red grouse, rock ptarmigan and mountain hares seen during transect walks in spring and summer, corroborated by total counts of all adults of these three species on defined study areas in spring and summer. AW did these counts for population studies of the three species (Watson *et al.* 1973, Watson *et al.* 1989, Watson *et al.* 1998, Watson *et al.* 2000), but the data have proved useful for the eagle study also. The distribution of rabbits was far more localised, but AW did total counts using dogs on some areas as well as counts on transect walks (Brown & Watson 1964), and made other estimates described in Watson *et al.* (1989).

Carrion from freshly dead red deer was available mainly in spring between late March and mid May (Watson 1971). That paper describes methods for counting the number of dead red deer seen during transect walks in spring. This was validated in 10 years against counts of total numbers of deer carcasses found by the deerstalkers on certain areas (Watson *et al.* 1989).

Watson *et al.* (1989) used the above data up to 1980 to provide estimates of the density of prey animals and of deer carcasses per given area. These were converted to mass per given area, using information on body weights. We continued the same methods for later years up to 1985.

Mean monthly temperatures are reported from daily values recorded at a climatological station at 339 m altitude at Braemar village (Meteorological Office Monthly Weather Report). Also from the same records we use the number of mornings with snow lying at Braemar (defined as snow covering at least half of the ground at 0900 hours) from October to May, and the number of days with air frost.

We transformed percentages by the angular transformation and all parameters apart from monthly mean temperatures (because there were minus

AW at a hide for photographing eagles at a nest in Deeside. AW helped zoologist Robert Carrick build the hide at this nest found by AW, and saw him into the hide for photography. (June 1950, Robert Carrick)

Golden eagle at the nest, with a large eaglet and prey, Deeside. (June 1950, Robert Carrick)

Robert was a lecturer at the University of Aberdeen and a brilliant field ornithologist. The eagle episode was to satisfy his keenness on bird photography. He subsequently moved to Canberra, Australia, where he was officer in charge at the Division of Wildlife Research (CSIRO), and was a founder of the Australian Bird Banding Scheme in 1953. By coincidences, SR now lives in Canberra and George Dunnet, who banded the first bird under that scheme, was Professor of Zoology at the University of Aberdeen when SR was a student.

values) by logarithms (ln). Hence we use Pearson correlations and linear regression. Because the number of adult pairs (and the number of birds) in consecutive years is likely to be autocorrelated, we also use proportionate change, expressed as the number in year 2 divided by the number in year 1. In a few cases where data could not be normalised, Spearman rank correlations were used.

RESULTS

ANNUAL NUMBER OF GOLDEN EAGLES
This comprised mainly the number of adult pairs, but in a few years immature birds colonised, later to become adult breeding pairs. In a few other years of population decline, AW judged that only one adult was present in each of certain home ranges. We therefore use two parameters: one, the annual number of adult pairs, and the second, the annual total number of adult birds either paired or single and immature birds holding a territory in spring and early summer.

Our analyses confirmed the earlier findings of Watson *et al.* (1989), now with a larger sample size due to five more years, which the number of pairs was correlated positively with the amount of carrion of red deer in the same spring (Table 3.1). The number of birds was also correlated with the amount of carrion, as well as the amount of mountain hares in spring (Appendix, Table 3.1.1).

The annual number of adult pairs and number of birds both showed a strong negative correlation with the calendar year, i.e. decreasing greatly in the course of the study (rs = −0.810 and −0.813, P <0.0001 and <0.0001).

NUMBER OF YOUNG REARED

Both the annual number of young reared per adult pair and the number of young reared per undisturbed adult pairs were correlated positively with the total amounts of prey (Table 3.2), the main effects being correlations with the amount of red grouse and rabbits. There were positive although weak associations with the amounts of ptarmigan and mountain hares, but there was a negative correlation with the amount of carrion.

A new finding was that the number of young reared per undisturbed adult pair was correlated negatively and significantly with the number of adult pairs and with the total number of birds in spring (i.e. with density: Table 3.3). This suggests the likelihood of a density-dependent regulatory effect, such that when density is high the birds breed less successfully.

NON-LAYING

This was negatively and significantly correlated with the amount of prey in the same spring, particularly with the amount of rabbits (Table 3.3). An expected result was that it was negatively correlated with the number of young reared per undisturbed adult pair.

Of further note was that breeding success was negatively correlated with the number of young reared per adult pair in the previous summer and the number of young reared per undisturbed adult pair in the previous summer (r = −0.317 and −0.434, P = 0.044 and 0.004). Hence, after good breeding success in the previous summer, there tended to be more non-laying in the following spring.

An eaglet looks out from its eyrie in Badenoch (July 1948). This nest cliff is the same one where H.B Macpherson photographed eagles as described in his book The Home-Life of a Golden Eagle *1910. Eagles still nest on this cliff.*

AMOUNT OF POTENTIAL FOOD

During the study period, there was a large negative correlation between the amount of food available as carrion (i.e. carcases of red deer) in relation to the calendar year (rs = –0.483, P = 0.0012), this declined sharply after 1964 and remained low. The change coincided with a much larger proportion of the deer population being shot.

The amount of prey showed a positive correlation with the calendar year, though not significant (rs = 0.290, P = 0.06). Total prey started at an extremely low trough during the mid 1940s, and although it fluctuated greatly from year to year, there was an obvious tendency to reach high points in the early 1960s and a peak in the early 1970s when there were several years with few deer carcases (Watson *et al.* 1989). During the early 1970s, numbers of all four main prey species were higher than usual. In particular, red grouse had a fairly long run of years with high abundance, falling to low numbers only in the late 1970s. Rabbits became more common than at any time since the arrival of myxomatosis and then maintained a higher abundance throughout the 1970s, rising again until 1980 and then maintaining the higher level.

The big fluctuations in numbers of red grouse and ptarmigan occurred over much of north-east Scotland, including our eagle study area. These fluctuations are cyclic, as defined by having periodicity more regular than random. There is evidence that these cyclic fluctuations are entrained by weather factors (i.e. fluctuations would occur for other inherent reasons within the populations, but the timing of the cyclic, fluctuations is influenced by weather (Watson *et al.* 1998, Watson *et al.* 2000, Watson & Moss 2008). It seems likely that there may be a similar wider influence on the fluctuations of mountain hares (Watson unpublished).

Rabbit numbers remained low for many years after the arrival of myxomatosis, but by the early 1970s some rabbits had become resistant to the virus and recovered from it, leading to a rise in abundance. Furthermore, the density of red deer showed a long-term increase, starting in the mid 1960s and continuing through the 1970s, 1980s and since then into the early 1990s. Their heavy grazing reduced the height of heather and increased the amount of bared trampled ground, already noticed by AW on our eagle study area in the early 1970s (Nethersole-Thompson & Watson 1974). These changes favoured rabbit grazing, colonisation and increase in numbers, and likely contributed to the observed increase of rabbits on the study area in and after 1980.

WEATHER

The annual amount of carrion in carcases of red deer during spring was negatively correlated with mean temperature in Braemar during January in the same year, and positively correlated with the number of mornings with snow at Braemar in that winter (Table 3.4). This, allied with other evidence on the condition of red deer in the area (Watson 1971), indicated that cold temperature and snow-lie increase the mortality of red deer in this high-lying snowy part of the Highlands. There was no relationship between the number of young eagles reared per adult pair or per undisturbed adult pair and the mean temperature at Braemar in any month from January to May. There was a negative correlation, however, between the number of April mornings when snow lay at Braemar and the number of young reared per adult pair as well as the number of young reared per undisturbed adult pair (n = 42, r = –0.351 and –0.364, P = 0.023 and 0.018).

Table 3.1. Annual number of adult pairs and birds, and proportionate change in their number in relation the annual estimated weights of prey and of carcases of red deer (n = 43 years for number of eagles, 42 for proportionate change in number).

	Prey		Carcasses		Total	
	r	P	r	P	r	P
Number of adult pairs	–0.260	0.10	**0.718**	**<0.0001**	0.695	<0.0001
Number of birds	–0.283	0. 07	**0.717**	**<0.0001**	0.689	<0.0001
Proportionate change in no. of pairs	0.094	0.55	–0.017	0.91	0.025	0.87
Proportionate change in no. of birds	0.017	0.91	**0.390**	**0.010**	0.444	0.003

The number of adult pairs, number of birds, and amount of carcases fell with the calendar year (rs = –0.810, –0.813 and –0.483, P <0.0001, <0.0001 and 0.0012), while the amount of prey rose with the calendar year though not significantly (rs = 0.290, P = 0.060).

Table 3.2. Aspects of breeding success in relation to number of adult pairs and of birds, i.e. density (n = 43 years).

	No. of adult pairs		No. of birds	
	r	P	r	P
No. of young reared/adult pair	–0.197	0.20	–0.283	0.067
No. of young reared/undisturbed adult pair	**–0.369**	**0.017**	**–0.420**	**<0.01**
Proportion of adult pairs non-laying	0.126	0.43	0.132	0.40

Table 3.3. Aspects of breeding success in relation to estimated weight of prey and carcases of red deer (n = 43 years).

	No. of young reared/adult pair		No. of young reared per undis- turbed adult pair		Proportion of adult pairs non-laying	
	r	P	r	P	r	P
Prey	0.388	0.011	0.515	0.0006	−0.392	0.011
Carcases	−0.15	0.34	−0.295	0.058	0.00	1.0
Total	−0.04	0.82	−0.16	0.30	−0.12	0.43
Red grouse	0.27	0.08	0.345	0.025	−0.28	0.07
Ptarmigan	0.13	0.39	0.02	0.90	0.08	0.63
Lagopus	0.317	0.039	0.341	0.027	−0.295	0.058
Mountain hare	0.05	0.74	0.19	0.24	−0.20	0.20
Rabbit	0.355	0.020	0.505	0.0008	−0.375	0.015

Table 3.4. Amounts of prey, carcases of red deer and total food in relation to weather at Braemar village in previous winter, mean air temperature in January, February, and number of mornings with snow lying (n = 42 years).

	January		February		Snow Mornings	
	r	P	r	P	r	P
Prey	0.287	0.065	0.02	0.90	−0.17	0.28
Carcases	−0.355	0.022	−0.26	0.09	0.437	0.004
Total food	−0.285	0.07	−0.25	0.11	0.349	0.024

DISCUSSION

Our results show that more eagles lived in the area when there was more carrion and they bred more successfully when there was more prey, as measured in one area over a long run of years. These results were similar to those found by Watson *et al.* (1992) for eagles in different study areas during a short run of four years. These results also support the correlation between the decline of eagle numbers in our area with less carrion per calendar year. Watson *et al.* (1989) presented a graph showing the decline of carcases, but did not compare the data against the calendar year. This was likely an effect of heavier shooting having compensated for deer mortality from starvation and cold during severe winters.

In the present study, we found a positive correlation between yearly breeding success of the golden eagle and yearly abundance of prey. Several workers have found such a relationship within areas, such as Tjernberg (1983) in Sweden, and Sulkava *et al.* (1999) in Finland. A study on Swedish mountain tundra (Nyström *et al.* 2006) revealed a strong positive correlation between breeding success and the yearly density index of the most important prey species, the two ptarmigan species (willow ptarmigan *Lagopus lagopus* and rock ptarmigan comprising 38% and 25% of the diet). In the northern Swedish boreal forest, Moss *et al.* (2012) found that reproductive performance was significantly correlated with the current year's indices of the primary prey populations of grouse species and mountain hares (in terms of an index of the annual population production and the percentage of territories with nestlings ≥4 weeks old).

During our study, more birds laid eggs when there was more food available, the proportion of non-laying being correlated significantly and negatively with the amount of prey in the same spring. To our knowledge, no other work has been done in Scotland on causes of non-laying in golden eagles. A failure to lay eggs in years of scarce prey was reported from Utah (Smith & Murphy 1979) and Sweden (Tjernberg 1983). A more recent study in Idaho by Steenhof *et al.* (1997) provided better evidence. They found that the proportion of pairs laying eggs was correlated positively with the abundance of jack-rabbits *Lepus californicus*, the main prey species, and negatively correlated with the severity of the previous winter. A similar correlation between non-laying and paucity of a main prey species has also been found in a study of the Spanish imperial eagle *Aquila adalberti*. The percentage of pairs that did not lay eggs increased during a period of years coinciding with rabbit haemorrhagic disease in most eagle home ranges, a disease that greatly reduced

A golden eagle chick ringed by AW at Mar, 1972.

The first Scottish golden eagle chick was ringed in 1923 and only a few or none were ringed in any year until the late 1950s. A total of 108 had been ringed by 1972 and about 50 have been ringed annually in the past several years. The first adult was not ringed until 1991. Up to 2017, 1,818 eagles have been ringed in the UK, including 30 adults (2017 British Trust for Ornithology data).

the abundance of rabbits, one of the main prey species (Margalida *et al.* 2007).

The findings of this long-term study of golden eagles in one Scottish area complement those of short-term studies in several areas (Watson *et al.* 1992), emphasising how eagle breeding density is related to the abundance of carrion in late winter and early spring; and their breeding success is correlated positively with the amount of prey available. However, the finding that breeding success was negatively correlated with the number of eagle pairs in the area was new. This result from one area might not be typical of all areas, but it could not have been detected by a study involving a short period even if several study areas were examined. This is due to potential confounding factors, differences between such study areas. Further long-term studies in several study areas would yield stronger evidence.

ACKNOWLEDGMENTS

We thank the Meteorological Office for data on snow-lie and other weather data.

REFERENCES

Brown, L.H. and Watson, A. (1964). *The golden eagle in relation to its food supply. Ibis 106,* 78–100.

Fasce, P., Fasce, L., Villers, A., Bergese, F. and Bretagnolle, V. (2011). *Long-term breeding density and density dependence in an increasing population of golden eagles Aquila chrysaetos. Ibis 153, 581–591.*

Margalida, A., Gonzalex, L.M., Sanchez, R., Oria, J., Prada, L., Caldera, J., Aranda, A. and Molina, J.I. (2007). *A long-term large-scale study of the breeding biology of the Spanish imperial eagle (Aquila adalberti). Journal für Ornithologie 148, 309–322.*

Moss, E.H.R., Hipkiss, T., Oskarsson, I., Häger, A., Eriksson, T., Nilsson, L.-E., Halling, S., Nilsson, P.-O. and Hörnfeldt, B. (2012). *Long-term study of reproductive performance in golden eagles in relation to food supply in boreal Sweden. Journal of Raptor Research 46, 248–257.*

Nethersole-Thompson, D. and Watson, A. (1974). *The Cairngorms: their natural history and scenery. Collins, London.*

Nyström, J., Ekenstedt, J., Angerbjörn, A., Thulin, L., Hellström, P. and Dalén, L. (2006). *Golden eagles on the Swedish mountain tundra – diet and breeding success in relation to prey fluctuations. Ornis Fennica 83, 145–152.*

Smith, D.G. and Murphy, J.R. (1979). *Breeding responses of raptors to jackrabbit density in the Great Basin desert of Utah. Raptor Research 13, 1–14.*

Steenhof, K., Kochert, M.N. & Mcdonald, T.L. (1997). *Interactive effects of prey and weather on golden eagle reproduction. Journal of Animal Ecology 66, 350–362.*

Sulkava, S, Huhtala, K., Rajala, P. and Tornberg, R. (1999). *Changes in the diet of the golden eagle Aquila chrysaetos and small game populations in Finland in 1957–96. Ornis Fennica 76, 1–16.*

Tjernberg, M. (1983). *Prey abundance and reproductive success of the golden eagle, Aquila chrysaetos, in Sweden.* Holarctic Ecology 6, 17–23.

Watson, A. (1957). *The breeding success of golden eagles in the north-east Highlands.* Scottish Naturalist 69, 153–169.

Watson, A (1971). *Climate and the antler-shedding and performance of red deer in north-east Scotland.* Journal of Applied Ecology 8, 53–67.

Watson, A (1982). *Work on golden eagle and peregrine in northeast Scotland in 1981.* Scottish Birds 12, 54-56.

Watson, A. and Hewson, R. (1973). *Population densities of mountain hares (Lepus timidus) on western Scottish and Irish moors and on Scottish hills.* Journal of Zoology 170, 151–159.

Watson, A., Hewson, R., Jenkins, D. and Parr, R. (1973). *Population densities of mountain hares compared with red grouse on Scottish moors.* Oikos 24, 225–230.

Watson, A. and Moss, R. (2008). *Grouse.* Collins New Naturalist 107, London.

Watson, A. Moss, R. and Rae, S. (1998). *Population dynamics of Scottish rock ptarmigan cycles.* Ecology 79, 1174–1192.

Watson, A, Moss, R. and Rothery, P (2000). *Weather and synchrony in 10-year population cycles of rock ptarmigan and red grouse in Scotland.* Ecology 81, 2126--2136.

Watson, A., Payne, A.G. and Rae, R. (1989). *Golden eagles Aquila chrysaetos: land use and food in northeast Scotland.* Ibis 131, 336–348.

Watson, A., Rae, S. and Payne, S. (2012). *Mirrored sequences of colonisation and abandonment by pairs of golden eagles Aquila chrysaetos.* Ornis Fennica 89, 229–232.

Watson, A. and Rothery, P. (1986). *Regularity in spacing of golden eagle Aquila chrysaetos nests used within years in north-east Scotland.* Ibis 128, 406–408.

Watson, J., Rae, S.R. and Stillman, R. (1992). *Nesting density and breeding success of golden eagles (Aquila chrysaetos) in relation to food supply in Scotland.* Journal of Animal Ecology 61, 543–550.

APPENDIX

Table 3.1.1. Stepwise multiple regressions for 1944–85 with the dependent variables: the number of eagle pairs, the proportionate change in that number, the number of young reared per undisturbed adult pair and the proportion of adults pairs considered not to have laid eggs. Independent variables were treated as the total weight (kg per km²) of all prey plus dead red deer, of prey plus dead red deer, of red grouse, of ptarmigan, of Lagopus, of rabbits, of mountain hares.

Dependent variable	Variables	T	P	VIF
Number of eagle pairs	Carrion (red deer)	4.50	0.0001	1.0
Change in eagle numbers	Mountain hares	2.40	0.0210	1.0
Young reared per undisturbed pair	Prey	5.13	<0.0001	1.6
..	Rabbit	3.09	0.0037	1.1
..	Red grouse	-2.15	0.0376	1.7
Proportion of pairs non-laying	Rabbit	-3.82	0.0005	1.5
..	Red grouse	-2.13	0.0394	1.2
..	Food total	-2.39	0.0219	1.3

This table shows the variable or variables that accounted for most of the variation in: a) the number of eagle pairs, b) the proportionate change in that number or c) the number of young reared per undisturbed adult pair and d) the proportion of adults pairs considered not to have laid eggs. In stepwise regressions, other variables drop out if of no significance. The adjusted R^2 for a), b), c) and d) combined was 0.315, 0.102, 0.460 and 0.231. In percentage terms these represent 5.6%, 3.2%, 6.8% and 4.8% of the variation being accounted for. Hence the bulk of the variation remains unaccounted for, due to other unmeasured factors. For a), all variables except dead deer dropped out because they were not significant, but ptarmigan and hares came close to being included ($T = -1.92$ and -1.91, $P = 0.062$ and 0.063). A small variance inflation factor (VIF), as in all those above, shows that an increase in variance of a regression coefficient due to correlation between independent variables is unimportant.

Analyses that use proportionate change are usually preferable to those using numbers alone, because the latter suffer from autocorrelation and consequent lack of full statistical independence. With golden eagles in this study area, however, only two of the 42 years with data on food involved a change in the number of pairs, so it seemed reasonable to present results using the number of pairs. When an analysis was done for proportionate change in the number of adult pairs, the only significant factor was the weight of mountain hares ($T = 2.40$, $P = 0.021$, VIF = 1.0, adjusted $R^2 = 0.102$ or in percentage terms only 3.2% of the variation accounted for by the weight of mountain hares).

4. NOTES ON BREEDING & FOOD OF GOLDEN EAGLES IN NORTH-EAST SCOTLAND

Adam Watson

SUMMARY

This short paper is presented to complement the previous chapter, adding details and brief analyses of breeding and food of golden eagles *Aquila chrysaetos* in a wider study area. The mean clutch size in 1944–59 was 2.0, and in 1960–1985 slightly smaller. In the former period, 83% of eggs in undisturbed nests hatched. Two cases of relaying were observed after first clutches failed. The main food at nests comprised red grouse *Lagopus lagopus scoticus*, ptarmigan *Lagopus muta*, mountain hares *Lepus timidus* and rabbits *Oryctolagus cuniculus*. After the disease myxomatosis severely reduced the number of rabbits, the proportion of rabbits taken to eagle nests fell, while that of other prey animals rose.

INTRODUCTION

The aims of this paper were to document the breeding success of golden eagles and food taken to their nests in a study area in north-east Scotland during 1943–85, which included the study area of Chapter 3. Breeding data were compared between the periods 1944–1959 and 1960–1985 as it was known that the number of breeding pairs of eagles in part of the study area increased during the former period and decreased in the latter (Watson *et al.* 2012). Data were noted in the course of work on numbers and breeding success of golden eagles in relation to food as described in Chapter 3 and previous studies (Watson 1957, Brown & Watson 1964, Watson & Rothery 1986, Watson *et al.* 1989).

In summer 1955, the virus disease myxomatosis arrived in the study area and killed almost all rabbits, apparently extirpating them completely in certain parts such as Glen Derry (Nethersole-Thompson & Watson 1974). Rabbits returned there in the early 1960s and 'in late summer and autumn are now back almost if not quite at their former abundance', but remained much scarcer in winter, spring and early summer than in years before myxomatosis. These large changes in abundance afforded an opportunity to study whether eagles would take a lower proportion of rabbits to their nests after the arrival of myxomatosis than before the virus came.

METHODS

During the breeding season I recorded fresh food at nests. Records of items in pellets are not presented, because not all food items within them could be readily identified in the field and no pellet contents were intensively analysed.

The frequency of visits was insufficient to determine reliable data on hatching success. Eggs that failed to hatch remained in nests for weeks, often until after the young fledged. However, occasionally an unhatched egg disappeared for unknown reasons. Also, if there had been two eggs and the nest later held one large chick but no unhatched egg, both eggs might have hatched and then one chick might have died, only to be eaten or removed. Hence values on hatching success are minimal.

RESULTS

TIMING OF BREEDING

There were few records of precise hatching or fledging dates and in many cases it was typical for eaglets to return to the nest after a first flight, irrespective of whether they had flown spontaneously or following disturbance by a person. They finally left after a few days of short return flights. The earliest record of hatching was a chick in a tree nest on 14 April 1944. The eaglet subsequently flew on 5 July. The sole egg in a tree nest of a different pair was chipping on 25 April 1949 and the chick flew on 26 June. In 1950, chicks hatched earlier than in any other year; in four cases where hatching was observed, two were in the third week of April and two in the early part of the last week of April. Two chicks at a high-lying crag nest had left the nest by 25 June 1950. A chick in each of two tree-nests in 1950 had left by 26 June, and a chick in another high-lying crag nest fledged on 28 June. There were too few records to test for any difference between hatching or fledging dates at crag or tree nests, as consideration of altitude and aspect would need to be considered as well as other possible influences. Generally, eaglets in north-east Scotland hatched in mid-late April and fledged in late June or early July.

CLUTCH SIZE AND HATCHING SUCCESS

Most clutches had two eggs. In 1945–59, before the numbers of pairs declined, few had one egg or three eggs, and the mean was 2.00 eggs (Table 4.1). Clutch size was slightly lower in 1960-81, with a mean of 1.89. Though small, this difference between means was statistically significant. In 1945–59, 83% of eggs hatched. In 1960–80 the sample size of six chicks hatching from 13 eggs was too small to justify analysis comparing the two periods. When data for both periods were combined the overall result was 74%.

RELAYING AFTER FAILURE

Only two cases of this were found, both in the same home range. In 1972, the two eggs had been broken and deserted between 2 and 16 April. On the latter date the

An eaglet lies on its eyrie with the remains of part-eaten red grouse. Most of our counts of food eaten by eagles have been of items found at or near nests during the breeding season, and this is important when studying breeding biology. Most evidence of food eaten at other times of the year has been from pellets found below eagle roosts and dissected to find what they had eaten.

The breast bone of a grouse lies below an eyrie with a characteristic wedge nipped out of it by an eagle's bill.

The remains of a red deer calf's hind leg (left) and that of a roe deer calf (right), collected from an eagle eyrie. The deer calves would have been dismembered before the parts were brought to the eyrie. A pellet cast by the young eagle in the eyrie lies to the lower right with the claws of a grouse protruding. And a second, very small pellet lies above that, composed of white feathers from an unidentified source.

A dead red deer calf at Derry Lodge wood, Mar, after an overnight snowstorm (15 April, 1951). The deer were fed in the area around the lodge in winter and spring, mostly with hay in the 1950s.

AW stands over the remains of a red deer calf (10 November, 1966). The carcass had been eaten by fox, golden eagle, chaffinch and coal tit.

The remains of a red deer stag that had died in winter lies below Coire na Poite of Carn a' Mhaim, Mar (June, 1951).

A red deer calf lies on the open hillside. Its hooves are still clean as it has not yet stood since being born. Newborn calves are very vulnerable to predation by eagles, but eagles can prey on them only for a few weeks in early summer each year. As soon as the calves are strong enough to follow on behind their mothers and mix with the herd there is less risk of them being attacked by an eagle.

A lamb sticks close to its mother. Ewes will fend off eagles from their lambs by rearing up on their hind legs while their lambs cower beneath them. Eagles have been known to take 1-3% of lambs in the Hebrides, and 0.15-2.4% can be expected to be lost to golden eagles in west Scotland overall (Campbell & Hartley 2004, Watson 2010).

Sheep carrion. The scattered white fleece from a sheep's body lying on dark heather-covered ground is obvious when walking transects and can be seen from hundreds of metres away. However, there is much less sheep carrion lying on the hills and moors of north-west Sutherland, as in other parts of the Highlands, since there has been a large reduction in the numbers of sheep grazed there in recent years.

hen was sitting on two warm eggs in a different nest, but both eggs were robbed between 17 April and the first week of May. In the first week of April 1978, two cold eggs lay on a nest, under a covering of great wood-rush. The hen moved to another nest and laid two eggs. On 24 June, the second nest contained one egg and one tiny chick, which flew in September. In both cases, human interference by the gamekeeper was suspected with the first clutch, but not proved.

FLEDGED YOUNG

Data on fledged brood size have been published elsewhere (Watson 1957, Brown & Watson 1964, and Watson et al. 1989). One eaglet was the commonest number of fledged young, but there were many cases of two, occasionally three.

Fledged young were observed to stay with parents usually for three months and to be fed by them during this period. I often heard eaglets away from the nest, yelping for food in July to October. The earliest date when I saw one catching a rabbit was in early October, and late October for catching a red grouse. In both cases the eaglet had fledged in early July. Hence, in cases where eaglets finally left the nest in late June, one would expect them to be able to catch prey at earlier dates.

I saw two cases where a juvenile eagle frequently associated with an adult pair through the following winter, spring and early summer. In both cases, the local stalker thought that the adult hen might still be feeding the juvenile in the following spring, because he often saw the two of them together, although he did not actually observe such feeding. In one of the two cases, the juvenile was not a strong flier. In both cases the adult hen partly built a nest in the following year but did not construct a deep lined cup, and no eggs were seen.

In a third case, a pair continued to associate with a juvenile through the first autumn and at least until 8 February the following year, 1970. On that date I saw the adult pair and an eagle in first-year plumage, all flying together while soaring above a hill. Each adult then dived separately at the eaglet, and the adult hen stretched out her talons at the end of her dive, almost touching the young bird's feet. Then all three flew high in a spiral, using a thermal current, and finally flew horizontally out of sight.

FOOD AT NESTS

This comprised mostly the same four species: red grouse *Lagopus lagopus scoticus*, ptarmigan *Lagopus muta*, mountain hare *Lepus timidus*, and rabbit, at all ranges (Table 4.2). However, the relative proportions at nests in each range varied in association with habitats. For instance, one range (4, Table 4.2) had hardly any alpine land, so the low proportion of ptarmigan in food items there was to be expected. Similarly, two ranges contained only small amounts of valley grassland, none of it on the lower glens, so rabbits were few, as reflected in food items at nests there. Eagles in range 3 during 1945 and 1946 fed their young overwhelmingly on rabbits, which comprised about 90% of food items taken to nests. In ranges 1–8, the proportion of rabbits in food items at nests underwent a huge decline after myxomatosis drastically reduced the abundance of rabbits. The fall in the proportion of rabbits compared with other food items was highly significant (Fisher exact test, $P < 0.0001$).

Table 4.1a. Percentage of golden eagle clutches observed with 1 or 2 or 3 eggs

	n	1	2	3	Mean	SE
1945-59	97	3	94	3	2.00	0.03
1960-81	90	14	82	3	1.89	0.04

Test for equality of variances $F = 2.68$, $P < 0.0001$ (hence variances unequal), test for difference between means, assuming that variances are unequal, $T = 2.22$, $P = 0.028$.

Table 4.1b. Hatching success of golden eagle clutches.

Clutch size (eggs)	1	2	2	3
No. hatched per clutch	1	1	2	3
No. of clutches	3	7	9	2
No. of chicks	3	7	18	6

Out of a total of 41 eggs observed in 1945–59, 34 hatched (83%). Far fewer were observed in 1960–81; one clutch of one egg produced no chick, three clutches of two eggs produced a chick each, and two clutches of three produced two chicks and one chick, i.e. a total of six hatching out of 13 eggs.

Young cuckoos are taken by golden eagles in Scotland in late spring. They are probably easily discovered when the foster parent meadow pipits are busy feeding them; the pipits make much noise calling as they fly to and fro, the cuckoos make a loud begging call and they have bright red gapes which they point skyward—eagleward.

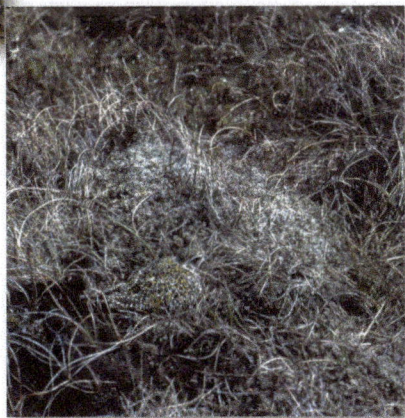

A golden plover, well concealed as it sits on its nest on the moorland. The bird is in the lower left centre of the photograph. This species is taken by golden eagles, although not frequently. The incubating birds sit quietly for about a month, offering few clues to predators such as eagles of their presence, and the off-duty plovers usually feed off the moorland on greener agricultural land. Like greenshank and dunlin, they only live in eagle country while nesting and rearing their chicks. So eagles are unlikely to be a great threat to these summer migrant waders.

Table 4.2. Percentage of food items on nests at ranges 1–8 in 1944–55 before myxomatosis reduced rabbit abundance, and 'After' in 1955–83, which were years after myxomatosis. 'Other' refers to 5 other Deeside ranges in 1944–55, and 'Angus' in 1957–80 to 6 ranges in the Angus glens: Clova, Esk and Isla, plus 2 in nearby north-east Perthshire. Because of small sample sizes within individual ranges, data from different ranges are lumped in each of the categories After, Other, and Angus.

	1	2	3	4	5	6	7	8	1-8	After	Other	Angus
Red grouse	30	37	18	36	32	32	32	38	31	39	33	50
Rabbit	20	12	57	28	20	7	20	17	24	3	23	7
Mountain hare	15	14	12	15	12	10	17	21	15	21	17	25
Ptarmigan	15	13	8	3	16	21	13	10	12	18	14	13
Red deer calf	5	2	0	3	4	4	0	3	2	1	2	4
Fox cub	5	10	2	0	4	7	3	3	4	4	2	0
Black grouse	5	0	0	3	0	0	0	3	1	2	2	0
Crow	5	0	2	3	4	0	3	0	2	1	2	0
Roe deer	5	0	0	3	4	0	3	0	1	1	2	0
Mole	0	2	2	3	0	4	4	3	2	1	2	0
Water vole	0	2	0	3	0	4	0	0	1	1	2	1
Brown rat	0	2	0	0	0	0	0	0	0.4	0	0	0
Stoat	0	0	0	3	0	0	0	0	0.4	0	0	0
Mallard	0	2	0	0	0	0	0	0	0.7	1	0	0
Golden plover	0	2	0	0	0	4	0	0	0.7	1	2	1
Trout	0	2	0	0	0	0	0	0	0.4	0	0	0
Adder	0	0	0	0	0	0	0	0	0	0	2	0
Ring ouzel	0	0	0	0	0	0	0	0	0	0	0	1
Lamb	0	0	0	0	0	0	0	0	0	0	0	2
Sheep	0	0	0	0	0	0	0	0	0	0	0	1
Red deer adult	0	0	0	0	0	0	0	0	0	0	0	1
Number of food items	20	49	51	39	25	28	30	28	271	84	66	98

A fulmar glides through the air. These birds are superb fliers but very ungainly on land when breeding and that is when the eagles catch them on the sea-cliffs. Their main defence is to spit a sticky oily fluid at an attacker. It is not known how golden eagles manage to catch them without becoming smeared with this oil, which can clog up an eagle's feathers and make flight impossible. Yet, they obviously can and do, because some eagles take them regularly, perhaps as their main prey item in summer.

DISCUSSION

The incubation period for golden eagles in Scotland is approximately six weeks, 43 days as recorded by Gordon (1955) and Watson (1997); the former time referred to a record supplied by AW from 1944. As such, the estimated laying dates for eagles in this study ranged between 3 and 14 March, calculated by counting days from the known hatch dates. This is earlier than that estimated for the eagles in the west Highlands, 21 March–1 April, and Scotland as a whole, 16 March–4 April (Watson 1997). In the west Highlands, the timing of egg-laying was later in years when the mean February air temperature was colder (Watson 1997). Yet in the early year of the north-east study, 1950, when young flew from several nests in late June, February was cold (mean air temperature 0.5° C at Braemar village, which is within the central part of the study area), but March very mild (5.2° C) and unusually snow-free even in glens above Braemar. The north-east study area was at higher altitude than that of Watson in the west Highlands, which extended down to sea-level, and Braemar lies at 339m. Therefore, as there was no clear effect of temperature, there is probably a less direct relationship between February temperatures and laying dates in north-east Scotland.

The few records in this study show higher or similar hatching success in north-east Scotland to that elsewhere. One study in Idaho recorded a hatching success of approx. 62% (Beecham & Kochert 1975) and another in Spain, 83% and 79% in the subsequent year (Sánchez-Zapata *et al.* 2000). In his book on golden eagle, Watson (1997, 2010) omitted hatching success, probably because this is difficult to ascertain, and few other studies of golden eagle have done so either. Yet, for example, this criterion was critical in the discovery of the detrimental effect of organochlorines

A mountain leveret cowers low among rocks and mossy hummocks. Leverets are left alone for most of the day, their mothers joining them briefly only to feed them. So the young animals' only protection from eagles is camouflage and remaining still so as not to betray their concealment.

A mountain hare grooms itself. Not all mountain hares turn completely white in winter, and as snow lie in Scotland can be so ephemeral, this is a suitable adaptation for camouflage, to hide from eagles. Pure white hares on a snow-free hillside would be obvious to an eagle, although hares will hide in hollows, among rocks or in holes in peat during the day.

A hen red grouse hen sits on her nest under cover of heather. The grouse's flecked russet, black and buff colouring mixes well in the shadows under the heath.

A hen red grouse and a chick hide in the heather. She is below left of centre and the chick is in the top right. Once hatched, red grouse pairs guard their chicks closely. On first sight of danger of an approaching predator they squat down into the heather and they all watch from their hiding places, waiting for the danger to pass. If the predator approaches too close, the adults jump up in alarm while the chicks take off in short rapid flight and land a short distance away. The adults continue to attract the predator's attention and lure it away. Once all is clear, the hen calls and gathers her chicks to safety.

A close up of the hen grouse in the former photograph

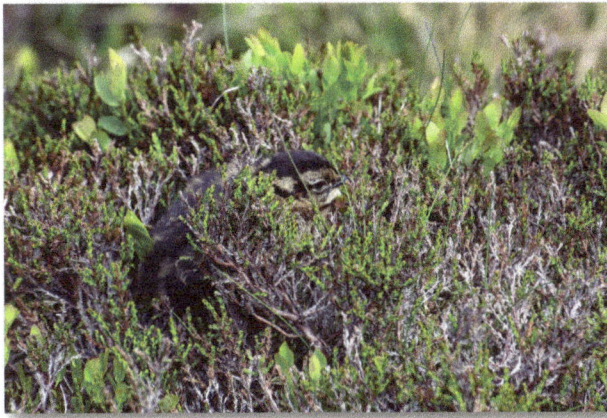

A close up of a red grouse chick in the same brood as in the above photograph.

The cock red grouse cock of the above family watches from his hiding place, ready to jump and distract any predator from the chicks.

A pair of ptarmigan in typical Scottish habitat for the species: short mountain heath with scattered boulders. They are well camouflaged and move slowly to avoid detection by eagles. The cock bird is on top of the large boulder on the left of the picture; the hen is between the boulders on the right.

A close-up of the above cock bird in the previous photograph. Cock ptarmigan spend much of their time in spring looking out over their territory, to defend it from other ptarmigan and to protect their hens. In so doing, they usually perch upon rocky outcrops or boulders, which are usually covered with grey lichens. Hence, the grey cock ptarmigan are well suited to blend in with their background.

A close-up of the hen ptarmigan in the above photograph. The hens spend most of their active time in spring feeding on mountain heath plants or sitting on their nests, which are hidden on the ground amongst such plants. So, their mixed colouring of different shades of browns, grey, black and white suits them better for camouflage than the predominant grey of the cock birds.

A scattering of white feathers where a ptarmigan has been plucked by an eagle and the body taken away to be eaten.

A hen ptarmigan on her nest next to rock, where her body outline is fused with that of the boulder, making her shape less obvious to any eagle high above. There was light rain when the photograph was taken. Water droplets hang from the plant stems and lie on her back, but she is unperturbed by such weather and her eggs would be warm beneath her insulating feathers.

The remains of a ptarmigan plucked by an eagle high on Glas Maol, found during a ptarmigan count (June, 1968). The dog is Dhubh, a cross English setter/English pointer.

on golden eagle breeding success, due to a high incidence of egg breakage (Lockie & Ratcliffe 1964).

Replacement clutches laid by golden eagles are rare (Watson 1997), and even if some went unrecorded, our study confirms this. One of only two such clutches found was laid within two weeks of the loss, thus concurring with Watson that such clutches must be laid soon after loss to allow sufficient time for the adults to rear the chicks.

At all home ranges where there were failures of breeding attempts, there was no known persecution. However, desertions at some eyries were known to have followed inadvertent disturbance by loggers, an artist, picnickers and others. Another effect might have been that unrecorded inadvertent disturbance may have occurred at nests where birds did not desert but were off their eggs long enough for an embryo to die, although this is speculation. However, as visitor numbers to the area have increased greatly since our study, it is likely that similar incidents could be more prevalent or become so. Intensive study is required to determine whether there is or might in future be such a detrimental effect on golden eagles.

The main prey species; red grouse, ptarmigan, mountain hare and rabbit, match those noted previously, (Watson 1957), and all are within the preferred weight range of 0.5–4 kg of all golden eagles (Watson 1997). Golden eagles elsewhere in Scotland take a wider range of prey species, with much local variation, and eagles are probably widespread across the Highlands because of this adaptability to local abundances of such prey. However, the abundance of such prey in the Highlands is influenced by human land uses, and rabbits, as an introduced species, are an example of this. The decline in the rabbit numbers was also caused by human intervention–an introduced virus. In New South Wales, Australia, the proportion of rabbits taken to nests by wedge-tailed eagles *Aquila audax* fell after an outbreak of rabbit calcivirus disease cut rabbit abundance, and the eagles switched to taking higher proportions of native species (Sharp *et al.* 2002). Wedge-tailed eagles in Western Australia ceased to breed after a major decline in the abundance of rabbits (Ridpath & Brooker 1986).

The present study has shown that the eagles in the north-east area were not dependent upon rabbits because they had alternative prey. However, the clutch sizes were smaller in the period post-myxomatosis when there were fewer rabbits. It is possible that there are breeding pairs of golden eagles elsewhere in Scotland which rely more heavily upon a narrow diet such as rabbits. A decrease in the availability of any such staple might reduce those eagles' breeding success. Therefore, if causes of any changes in the status of breeding eagles are to be fully understood, it is important that not only breeding data be recorded, but also foods eaten.

REFERENCES

Beecham, J.J. and Kochert, M.N. (1975). Breeding biology of the golden eagle in southwestern Idaho. The Wilson Bulletin, 506–513.

Brown, L. H. and Watson, A. (1964). The golden eagle in relation to its food supply. Ibis 106, 78–100.

Campbell, S., & Hartley, G., 2004. Investigation into golden eagle predation of lambs on Benbecula in 2003. SASA.

Gordon, S. (1955). The golden eagle. Collins, London.

Lockie, J. D., and Ratcliffe, D. A. (1964). Insecticides and Scottish golden eagles. British Birds 57, 89–102.

Nethersole-Thompson, D. and Watson, A. (1974). The Cairngorms: their natural history and scenery. Collins, London.

Ridpath, M.G. and Brooker, M.G. (1986). The breeding of the wedge-tailed eagle Aquila audax in relation to its food supply in arid Western Australia. Ibis 128, 177–194.

Sánchez-Zapata, J.A., Calvo, J.F., Carrete, M. and Martínez, J.E., 2000. *Age and breeding success of a golden eagle Aquila chrysaetos population in southeastern Spain. Bird Study* 47, 235–237.

Sharp, A., Gibson, L., Norton, M., Ryan, B., Marks, A. and Semeraro, L. (2002). *The breeding season diet of wedge-tailed eagles (Aquila audax) in western New South Wales and the influence of rabbit calicivirus disease. Wildlife Research* 29, 175–184.

Watson, A. (1957). *The breeding success of golden eagles in the north-east Highlands. Scottish Naturalist* 69, 153–169.

Watson, A. and Rothery, P. (1986). *Regularity in spacing of golden eagle Aquila chrysaetos nests used within years in northeast Scotland. Ibis* 128, 406–408.

Watson, A., Payne, A.G. and Rae, R. (1989). *Golden eagles Aquila chrysaetos: land use and food in northeast Scotland. Ibis* 131, 336–348.

Watson, A, Rae, S. and Payne, S. (2012). *Mirrored sequences of colonisation and abandonment by pairs of golden eagles Aquila chrysaetos. Ornis Fennica* 89, 229–232.

Watson, J. (1997). *The golden eagle. Poyser, London.*

Watson, J. (2010). *The golden eagle, second edition. Poyser, London.*

FOOTNOTE

Studies of pellets exaggerate the proportion of smaller animals eaten whole, whereas observations of food items that include bones too big for eagles to eat may exaggerate the proportion of large animals eaten (Lockie & Ratcliffe 1964). One problem with using pellets for analysis of food eaten was that most remains of red grouse and ptarmigan did not differ enough to distinguish the species for certain, especially given that white feathers occur in both. Another was that AW could easily tell the dark grey-blue fur from the upper-parts of adult mountain hares, but could not distinguish the fur of rabbits from that of small leverets. Therefore, data for red grouse and ptarmigan were combined, and likewise the data for mountain hare and rabbit. The toe-nails and beaks of red grouse and ptarmigan were useful clues, and the hard thick skin on the undersides of the toes was often undigested. Small pieces of food from the crops of killed red grouse and ptarmigan often appeared, totally undigested by the eagle and clearly recognisable as to plant species. Though interesting, this was seldom an aid to telling the species, because red grouse and ptarmigan eat much heather *Calluna vulgaris* and blaeberry *Vaccinium myrtillus*, and the tell-tale occurrence of alpine specialists such as least willow was too infrequent in the eagle pellets to be of much use.

In three cases AW observed an eagle catching a ptarmigan, and four cases of killing a red grouse, four for mountain hares, five for rabbits, and seven cases of feeding on carcases of red deer *Cervus elaphus*. These are very small sample sizes, despite many years with field observations. Adult golden eagles spend a small proportion of the day killing prey, and so are seldom seen to catch prey or eat food. In ranges 1-8, red grouse and ptarmigan combined formed 47% of items identified in 109 pellets before myxomatosis, while hares and rabbits formed 39%. In 68 pellets after myxomatosis, the proportions were 56% and 32% respectively.

5. COMPARISONS OF GOLDEN EAGLE HOME RANGE OCCUPANCY & BREEDING SUCCESS IN NORTH-WEST SUTHERLAND BETWEEN 1957-1960 & 1982-1985

Stuart Rae & Adam Watson

SUMMARY

Occupancy and breeding success at 16 golden eagle home ranges in a north-west Sutherland study area were compared between 1957–60 and 1982–85. Home-range occupancy in 1982–85 was 25% less than in 1957–60. A higher proportion of eagles failed to rear young due to not laying or breaking eggs in 1957–60 than in 1982–85, which coincided with the time of maximum effects of organochlorine compounds on eagles in the 1950s and early 1960s. Although a higher proportion of eagle pairs laid eggs in 1982–85, they did not rear more young than in 1957–60. More nesting eagles were disturbed and more pairs failed to rear young from eggs laid in the later period.

INTRODUCTION

Two studies of breeding behaviour of golden eagles *Aquila chrysaetos* in several areas of the Scottish Highlands in 1957–60 and 1982–85 compared the nesting densities and breeding successes of eagles between regions and habitat types in the Scottish Highlands (Brown & Watson 1964, Watson *et al.* 1992). However, although part of these studies was in the same area of north-west Sutherland in two distinct time periods, the results were not tested for differences between them over time. Therefore, the aim of the present study is to complement those studies by comparing the home-range occupancy and breeding success of golden eagles within those ranges in north-west Sutherland between the two study periods.

The study area ranges in height from sea level to hills over 900m. Most of the intervening ground is undulating peaty moorland and steep-sided valleys, and the annual precipitation varies greatly between the low-lying coastal area and among the high hills. The vegetation on the lower ground is dominated by heather *Calluna vulgaris*, grasses and sedges. There are only small stands of woodland, and arctic-alpine vegetation is prevalent above 500m. Although greatly reduced since the studies, sheep grazing was one of the main land uses, together with deer stalking, tourism and salmon fishing. Red deer *Cervus elaphus* and domestic sheep *Ovis aries* are the main large herbivores in the area and their carrion is food for eagles. Other food species are rock ptarmigan *Lagopus muta* on the summits and ridges, red grouse *Lagopus lagopus* on the heather-rich moors, and mountain hares *Lepus timidus* on both. There are rabbits *Oryctolagus cuniculus* in some lower, grassy areas.

METHODS

The golden eagle home ranges used for this study lay within both regions studied in 1957–60 and 1982–85 (Brown & Watson 1964, Watson *et al.* 1992), although not all home ranges were monitored in both studies. Locations of some nest sites in the study area were accrued by Leslie Brown from his own searches or via information from local people and visiting ornithologists. These were all checked, and all other potential nest sites were searched for occupied and unoccupied nests in 1957–60 by Brown & Watson (1964). Most nests were found in their first year of use during the study, and nests were monitored annually. Golden eagles tend to use only one of several in a cluster and clusters of nests were distinct for each home range (Watson 1997). A home range was defined as an area where a pair of eagles lived in a cluster of nest sites, within which only one nest was used per annum (Watson *et al.* 2012). Known nest sites were again checked for occupancy in 1982–85, as well as any other potential nest cliffs within the original area and in the extended later study area (Watson *et al.* 1992). The detailed methods for the eagle surveys were as later used in national surveys of golden eagles (Dennis *et al.* 1984, Hardey *et al.* 2006).

Breeding success was found by repeated visits to the nest sites. Categories were: nest occupied, eggs laid, young hatched and number of young reared (Watson *et al.* 1992). Exact clutch sizes were seldom recorded as nests were viewed from a distance to minimise disturbance, although some nests were accessed; e.g. if easily approached or to determine cause of failure, to ring chicks, or to note food brought to the nest. Not all breeding attempts were fully and comprehensively monitored to provide all data for all home ranges in all years. Therefore the data sets used for analysis of the effects below are based on samples which varied in number.

Weather can interact with the availability of prey to reduce golden eagle breeding success (Steenhof *et al.* 1997). However, it was not uniform over the whole area at all times. The weather varied between low-level, coastal home ranges which tended to have more sunshine, and higher, farther inland ranges which tended to have more rain. It was considered inappropriate to try to analyse some possible effects of weather on breeding success because weather was not recorded at nest sites, nor at any potential hunting areas assessed for prey. Data were collected for four years in each of the two study periods to allow for any within-period variability (e.g. caused by local weather at each nest site). Two-tailed t-tests were used for differences between the two sample groups. Data for 1967 were supplied by Leslie Brown and these have been added to the graphs for visual comparison, but they were not used in statistical analyses.

Variables compared between the two study periods were: 1) home-range occupancy, which was defined as a pair present and a nest built; and 2) breeding success, which was analysed in terms of reasons for failure (viz.: eggs robbed, human disturbance, non-laying of eggs, eggs broken, eggs failed to hatch, and failure to rear young). Egg collection was a serious problem in the UK in the 1950s and 1960s, and might still have been so in the 1980s. We also considered other possible effects of human disturbances including deliberate or unintentional destruction of nests or eggs (such as nests burnt), birds disturbed (so eggs or small young are lost), and eggs robbed.

Eagles in the study area were known to eat sheep carrion (Brown & Watson 1964, Watson *et al.* 1992), and breeding failure due to non-laying of eggs and broken eggs was a known detrimental effect of organochlorine compounds, such as dichlor-di-phenyl-trichlor-ethane (DDT) or Dieldrin, prior to its ban as a sheep dip in 1965 (Lockie & Ratcliffe 1964, Lockie *et al.* 1969). We tested for differences in reproductive parameters between study periods which were within the periods of use and non-use of these chemicals. A final category (failure of undisturbed pairs) was used as a comparative measure of natural causes of breeding failure.

RESULTS

HOME-RANGE OCCUPANCY
Sixteen home ranges were monitored in both study periods, 1957–60 and 1982–85. One home range identified prior to the study was unoccupied in all years, so it was excluded from all further analyses. One home range was unoccupied in three of four years in the earlier study period and unoccupied in all years in the second period. One home range was unoccupied in three years, and four in all years in the later period. These four were unoccupied for more than a three year period, so they were assessed as abandoned (Whitfield *et al.* 2007). This was a loss of 4 from 16 pairs, 25% of the number of breeding golden eagle pairs in the area between 1957 and 1985. If the pair that was lost prior to the study is included, the loss would be 5 from 17 pairs, 29% of the population in the earlier 1950s.

In 1957–60, a minimum of 15 home ranges were recorded as occupied in any one year with a mean occupancy rate of 90.2% ± standard error (SE) 3.30, (range 85.5–100%, n = 7, 11, 8 & 8). In 1982–85, 12 home ranges were occupied in any one year with a mean occupancy rate of 69.8% ± SE 1.80 (range 66.7–75%, n = 15,16,16,16) (Figure 5.1). There were significantly fewer home ranges occupied in the 1982–5 period compared with those in 1957–60 (t = 5.42, P = 0.002). In 1967, there were 11 occupied home ranges from a sample of 12 sites (91.7%). The single unoccupied range was subsequently unoccupied in 1982–85.

Cloud lies along the ridge of Foinaven, one of the highest hills (914 m) in Sutherland. The ridge lies along a south-east to north-west line, across that of the prevailing south-westerly winds. In Scotland, these are moisture-bearing winds, and the high central part of the north-west Sutherland study area has much higher precipitation than the coastal moorlands.

The Moine, an extensive area of wet heath and blanket mire on the moorland in north-west Sutherland. Purple moor grass is the dominant plant, with heather and cotton grass other constant species. Red deer graze here and there are low densities of red grouse and mountain hares. The high hill is Ben Hope (927 m).

Most human settlements in north-west Sutherland are near the coast or in the larger glens, like this old castle atop a coastal cliff. This pattern of settlement has left large tracts of country quiet enough for golden eagles to live in, although hill-walking has increased greatly since 1958 and the hills are not as quiet as they were then. Hills such as Ben Loyal (765 m), shown here, might not be especially high, but their rugged form that attracts many visitors.

The rocky summit ridge of Foinaven, ptarmigan country. Few other birds or mammals subsist here; mountain hares are sparse and meadow pipits live there only during the short spring/summer season. Golden eagles are perhaps the most likely birds to be seen here in winter.

Much of the vegetation on the high stony ridges is out of reach of red deer, the main wild herbivore in the area, as they seem to avoid walking across steep loose boulder fields. Hence, there are lush beds of such food plants as blaeberry and crowberry for the ptarmigan and mountain hares to feed on.

There are few roads in the Sutherland study area and many of them are single-track. However, there have been considerable upgrades of the roads and much greater usage between 1958 and 2018. Tourism, often car-based, is a major industry in the area. The area is not as remote and unfrequented as it once was.

The quartzite-topped mountains of Arkle and Foinaven stand high above the lower undulating land formed from Lewisian gneiss. The many lochans (small lakes), heathery ridges and rocky knolls are characteristic features of this lower ground and hunted over by golden eagles looking for red grouse and mountain hares.

Spring snow can fall on the high ground of Sutherland, as on all the Scottish hills, although it seldom lies long. Even in winter, most of the tops are generally snow-free. Although ptarmigan turn white in winter, or at least partially so, they do not need snow to hide from eagles. They hide close by rocks and in boulder fields, and quartzite is as white as snow.

A shower passes across the Sutherland moorland. Such a shower would have little effect on a golden eagle. They can sit out a day, probably even a few days of rain, but prolonged spells of heavy rain probably curtail their hunting ability. If this were to happen when they have chicks, it might reduce their breeding success.

A fire runs across the Sutherland moorland; several square kilometres were burnt out by this fire that began at the side of a road. This was in May ,when many moorland birds were nesting, and whether it was lit deliberately or accidentally, it must have killed much wildlife.

The nearly burnt-out nest of a pair of greenshanks. A greenshank had been sitting on these eggs the day before the fire shown above. The nest was not burnt, but the eggs were subsequently taken. So little green ground was left around it that after the fire, crows, which were seen hunting in the area, probably soon found the eggs.

Looking westwards at sunset from north-west Sutherland, the sun slips behind An Garbh-eilean and the north Atlantic ocean. A red sky at night, an eagle's delight. The clear red sky is a sign of settled weather in the west, the direction from whence the prevailing winds bring rain. No heavy clouds—no rain to come tomorrow, so a clear day for the eagles to hunt.

The 190 m high cliffs of Clo Mor. Golden eagles hunt along these cliffs for fulmars, and as recently proved by remains found at an eyrie several kilometres away, they also take puffins.

A man turns hay in a field on a narrow strip of cultivated land on the Sutherland coast at Badcall, on the Rhiconich-Kinlochbervie road (September 1958). Much of the Sutherland land is like this, with moorland stretching from the headwall stone-dykes of the inby land to the foot of the high mountains, as in the background of this view. The high ridgeline of Foinaven is clear in sunshine on the left and Arkle on the right edge.

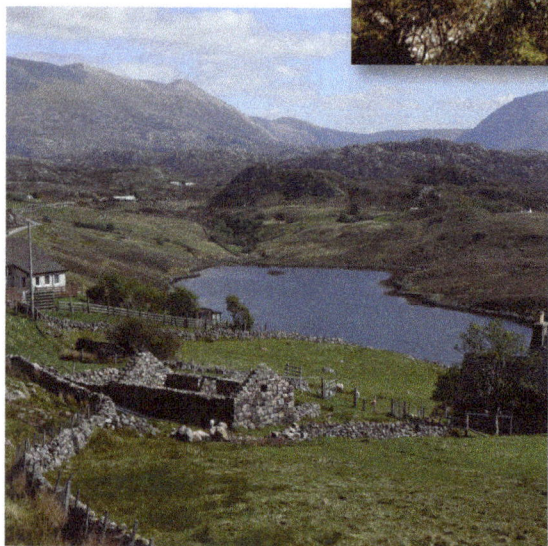

A retake of the same view (May 2018). There is no hay being grown in any of the fields, and only a few sheep graze in one of them. The iron roof on the building in the early photograph has gone in 2018, a new house has been built on the left and the house on the right edge has been modernised. There is no change to the land outside the headwall.

BREEDING SUCCESS

Breeding success of all golden eagle pairs in the study area in all years was analysed for two criteria, the total number of young reared per pair and the total number of young reared per successful pair. There was no difference in the numbers of young reared per breeding pair of golden eagles between 1957–60 and 1982–85 (t = 0.96, P = 0.37). The mean number reared in 1957–60 was 0.49 ± 0.10, range 0.2–0.64 (n = 5,11,7,7) and in 1982–85 it was 0.45 ± 0.32, range 0.36-0.5 (n = 10,11,11,12) (Figure 5.2). There were 0.4 young reared per pair in 1967. Although there were two pairs of sibling young reared in the earlier period, there was no significant difference between the number of young reared per successful pair of breeding golden eagles in 1957–60 than in 1982–85 (t = 1.67, P = 0.15). The mean number reared in 1957–60 was 1.1 ± 0.70, range 1.0–1.3, and in 1982–85 it was 1.0 ± 0.0, range 1.0–1.0. There was one chick reared per successful pair in 1967.

More pairs of golden eagles did not lay eggs in 1957–60 than in 1982–85 (t = 2.67, P = 0.03). In 1957–60, the mean proportion of pairs that did not lay eggs was 34.4% ± SE5.41, range 27–50% (n = 6,11,7,6), in 1982–85, the mean proportion of pairs that did not lay eggs was 13.2% ± SE5.77, range 0–27.2% (n = 10,11,11,12). The figure for this in 1967 was 10% (n = 10). We seldom knew why eggs did not hatch, or if all eggs in a clutch hatched. Therefore any figures for the number of clutches where eggs were broken would be a minimum figure.

The true figure could and would probably have been higher if there was an effect of organochlorine compounds in those eggs in the earlier period. However, from a sample of 20 clutches that were laid in 1957–60, there were two clutches known to have been broken (10%), compared with only one clutch in a sample of 32 (3.1%) in 1982–5. These data were too few for separate analyses so, because non-laying and egg breakage are known effects of organochlorines in the eagles' diet, these were added to the proportion of non-laying for analysis of possible effects of organochlorines on breeding eagles. More golden eagle pairs did not lay or their eggs were broken in 1957–60 than in 1982–85 (t = 2.78, P = 0.03). In 1957–60, the mean proportion of pairs that did not lay or broke eggs was 42.5% ± se 8.68, range 27.2–66.7% (n = 6,11,7,6), in 1982–85, the mean proportion was 15.7% ±SE 4.18, range 10–27.2% (n = 10,11,11,12).

HUMAN DISTURBANCE

In 1957–60, the mean proportion of clutches robbed was 5.82% ± SE 3.35, range 0–14.3% (n = 6,11,7, 7) and in 1982–85, the mean proportion was 7.25% ± se 4.75, range 0–20% (n = 10,11,11,12). There was no significant difference between the proportions of clutches taken by egg collectors in 1957–60 and 1982–85 ((t = 0.24, P = 0.82). Although there was only one annual record from the time between the study periods, when 31% (n = 11) were taken in 1967, this was at a higher level than during the two main study periods. There were more breeding failures due to any human disturbances of nesting golden eagles in 1982–85 than in 1957–60 (t = 4.08, P< 0.01). The mean proportion of breeding attempts that failed due to any human disturbance in 1957–60, was 5.82% ± SE 3.35, range 0–14.3% (n = 6,11,7,7) and in 1982–85, the mean proportion was 22.62% ± SE 2.12, range 18.2–27.3% (n = 10,11,11,12) (Figure 5.3).

The figure for 1967 was relatively high, all due to eggs having been robbed. There was no significant difference between the proportions of clutches robbed between the two main study periods, therefore one or more other human disturbances were probably greater causes of breeding failure in 1982–85, although no single type of disturbance could be determined as a main cause. The analyses of breeding success that follows were for data from those breeding attempts that were not influenced by human disturbances.

BREEDING SUCCESS BY PAIRS OF GOLDEN EAGLES THAT WERE NOT DISTURBED

There were no breeding attempts where two young eagles fledged in 1982–85, therefore the proportion of pairs that reared young is the measurement subsequently used, and these proportions in 1957–60 and 1982–85 were similar (t = 1.16, P = 0.29). In 1957–60, the mean proportion of pairs that reared young was 42.2% ± SE 13.7, range 16.7–60% (n = 6,10,7,6), in 1982–85, the mean proportion was 59.0% ± SE 4.98, range 44.4–66.7 (n = 8,8,9,9). The figure for 1967 was 57.1% (n = 7).

Some pairs of eagles did not lay eggs, and of those that did lay, more pairs failed to rear young in 1982–85 than in 1957–60 (t = 3.4, P = 0.02). In 1957–60, the mean proportion of undisturbed eagle pairs that failed to rear young from eggs laid, attributable to probable natural causes was 4.73% ± SE 4.73, range 0–14.2% (n = 7, 4 & 3), and in 1982–85, the mean proportion was 26.1% ± SE 4.13, range 14.2–33.3% (n = 7,7,6,7) (Figure 5.4). The figure for 1967 was 20% (n = 5).

DISCUSSION

The main findings of this study were that there were fewer breeding pairs of golden eagles in north-west Sutherland in the 1980s than in the late 1950s and early 1960s; and their breeding success was no better in the 1980s than in the earlier period, despite the likely effects of organochlorine compounds DDT or Dieldrin prior to 1966.

It seems that any recovery after 1966 from the probable negative effect of DDT/Dieldrin on the breeding success of eagles in the 1957–60 period was offset in 1982–85 by more disturbance of breeding birds and fewer pairs rearing young due to natural causes. Natural causes include weather or the availability of food. Because the study was over four years in each period, reduced productivity is more probably due to less available food (or a particular type of food) in the 1980s. The apparent high impact of egg theft between 1960 and 1982 seems to have justified the secrecy of Brown & Watson (1964), because despite their reluctance to divulge exactly where there were so many golden eagle nests, many clutches were still robbed, particularly in 1967. Hopefully, this fashion has largely passed, although it is known to persist and cannot be ignored as a negative effect on eagles in Scotland.

The effects of organochlorine compounds were probably not a cause of lower breeding success in 1982–85, because there were fewer incidents of non-laying or egg breakage. These chemicals were no longer used in sheep dips, so should not have been ingested by eagles eating sheep carrion. However, the diet of some eagles in the area, especially those in home ranges near the coast, includes fulmar *Fulmarus glacialis*, a marine species. Golden eagles on Rum in western Scotland, which had low breeding success, had high levels of DDE (the in-body metabolite of DDT) polychlorinated biphenyls (PCBs) and mercury, which had probably been ingested with a high proportion of seabirds in their diet (Furness *et al.* 1989). Golden eagle eggs from western Scotland where seabirds were frequently eaten also contained high levels of

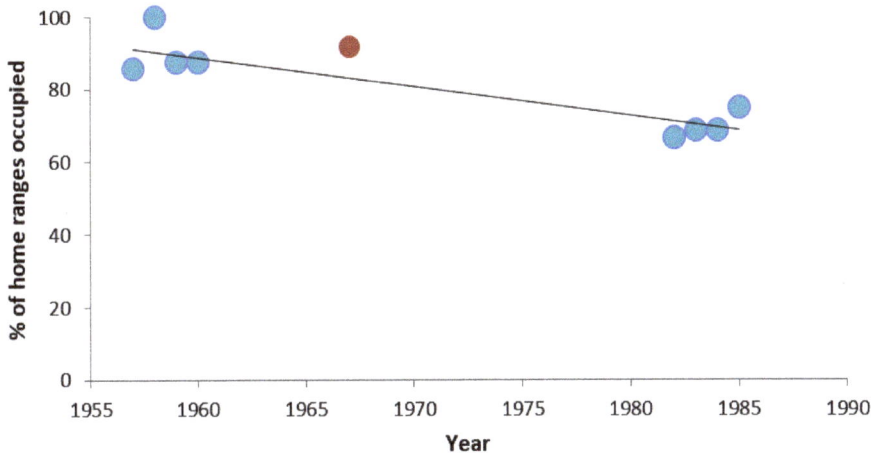

Figure 5.1. *The proportions of golden eagle home ranges that were occupied in the north-west Sutherland study area in 1957–60, 1967 and 1982–85.*

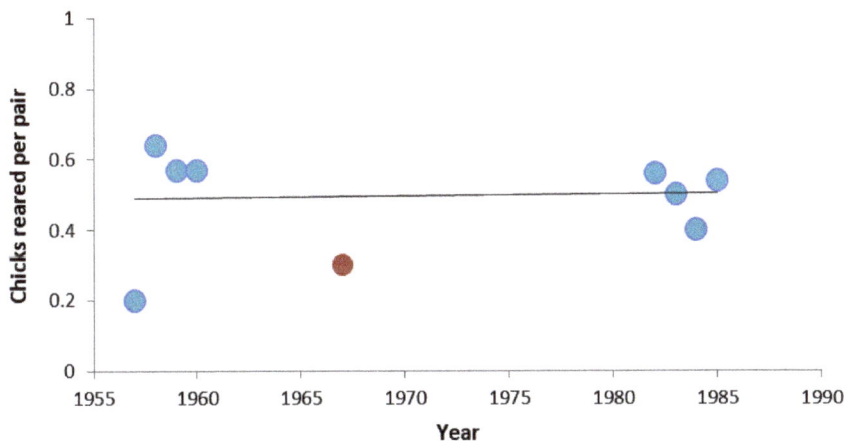

Figure 5.2. *The mean numbers of young reared per breeding pair of golden eagles in north-west Sutherland in 1957–60, 1967 and 1982–85.*

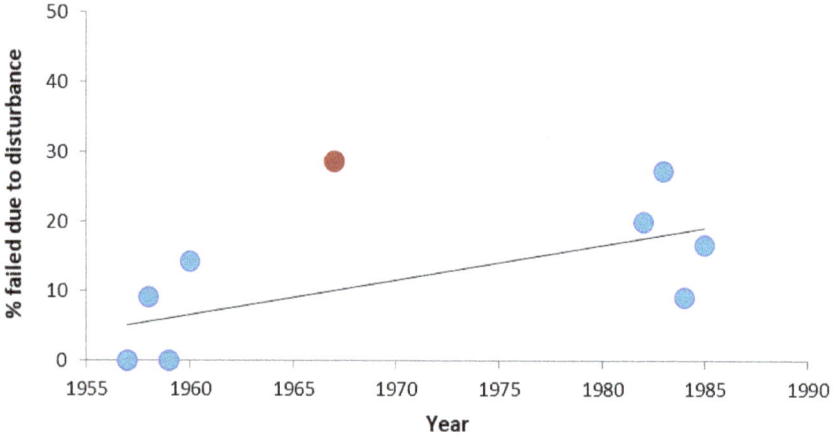

Figure 5.3. The annual proportions of breeding attempts by golden eagle pairs that failed due to human disturbance in north-west Sutherland in 1957–60, 1967 and 1982–85.

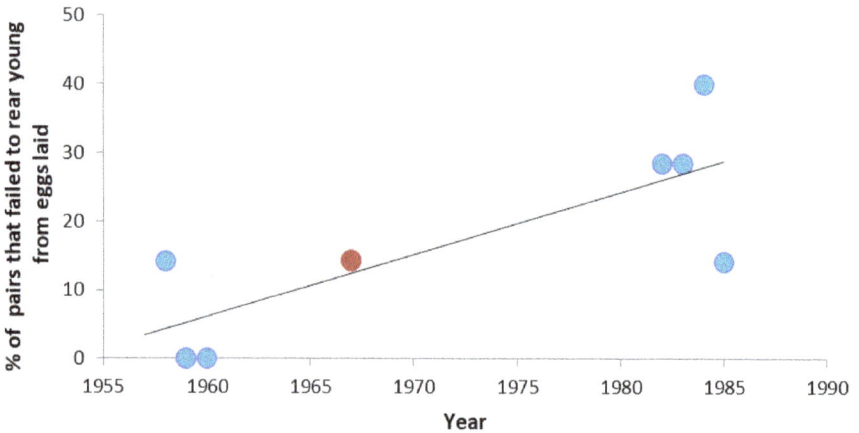

Figure 5.4. The annual proportions of undisturbed golden eagle pairs that laid eggs then failed to rear young in 1957–60, 1967 and 1982–85.

organochlorines (Newton & Galbraith 1991). There has been similarly lower breeding success in golden eagles in western Norway as a result of eggshell-thinning and breakage, which has been linked to organochlorine load caused by intake of marine species such as fulmar (Nygård & Gjershaug 2001). Therefore, there might still be some chemical contamination of golden eagles, especially in coastal home ranges, and this should perhaps be monitored more closely.

Disturbance, whether intentional or non-intentional, can cause eagle pairs to fail in a breeding attempt, and loss of golden eagle habitat can result from increases in human encroachment such as roads or recreation (Watson & Dennis 1992). Disturbance was a cause for failure in the 1980s, and it might also have been a cause for the abandonment of some home ranges since the 1950s. Accessibility of nest sites is a factor in the assessment of disturbance of golden eagle nest sites (Brown 1969) and the possible effect of these factors on occupancy and breeding success will be investigated in a future study. Home ranges where golden eagles have poor breeding success are also probably more likely to be abandoned than those with a history of good breeding success (Whitfield *et al.* 2007), and variation in breeding success in several regions of the Highlands has been found to be related to differences in the abundance of prey (Brown & Watson 1964, Watson *et al.* 1992). There were lower estimates for food availability in the study area in the 1980s than in the 1960s (Rae & Watson–Chapter 7). Therefore lower productivity in the 1980s might have been an effect of less available food than in 1957–60. Further study will test whether a difference in the availability of food within the same area over later years has had any effect on breeding success.

Three of the main negative factors on breeding in golden eagles in north-west Sutherland have been non-laying in the 1957–60 period, more disturbance and fewer young reared in 1982–85. The first effect, which was probably caused by contamination with organochlorine compounds DDT or Dieldrin, and the second were both anthropogenic. The third factor was likely caused by a reduction in the amount of food available to the eagles. If food availability were the cause, it could also be partly anthropogenic, because human land use can affect abundance of food for eagles, via land management practices such as sheep husbandry and deer culling. In conclusion, there were noticeable individual effects of land use, and cumulative changes in human land use probably produce the greatest impact on golden eagles in their shared habitat.

ACKNOWLEDGMENTS

Leslie Brown was a pioneer of this study, whose unpublished notes greatly enhanced the dataset and our expanded knowledge of eagles in the area. Jeff Watson led the study in 1982-85 under the administration of the Nature Conservancy/Nature Conservancy Council. We also acknowledge useful observations by David Jenkins and Derek Ratcliffe for finding some pairs in earlier years, back to 1937, and several shepherds and deerstalkers for local information.

REFERENCES

Brown, L. H. (1969). *Status and breeding success of golden eagles in north-west Sutherland in 1967. British Birds 6, 345–363.*

Brown, L. H. and Watson, A. (1964). *The golden eagle in relation to its food supply. Ibis 106, 78–100.*

Dennis, R.H., Ellis, P.M., Broad, R.A. and Langslow, D.R. (1984). *The status of the golden eagle in Britain. British Birds 77, 592–607.*

Furness, R. W., Johnston, J.L., Love, J.A. and Thompson, D.R. (1989). *Pollutant burdens and reproductive success of golden eagles Aquila chrysaetos exploiting marine and terrestrial food webs in Scotland. In Meyburg, B.U. and Chancellor, R.D. (eds.). Raptors in the modern world, 495–500. WWGBP, Berlin, London & Paris.*

Hardey, J., Crick, H.Q.P., Wernham, C.V., Riley, H.T., Etheridge, B. and Thompson, D.B.A. (2006). *Raptors: a field guide to survey and monitoring. The Stationery Office, Edinburgh.*

Lockie, J. D. and Ratcliffe, D. A. (1964). *Insecticides and Scottish golden eagles. British Birds 57, 89–102.*

Lockie, J. D., Ratcliffe, D. A. and Balharry, R. (1969). *Breeding success and organo-chlorine residues in golden eagles in west Scotland. Journal of Applied Ecology 6, 381–389.*

Newton, I. and Galbraith, E.A. (1991).*Organochlorines and mercury in the eggs of golden eagles Aquila chrysaetos from Scotland. Ibis 133, 115–120.*

Nygård, T. and Gjershaug, J O. (2001). *The effects of low levels of pollutants on the reproduction of golden eagles in Western Norway. Ecotoxicology 10, 285–290.*

Steenhof, K., Kochert, M. N. and McDonald, T. L. (1997). *Interactive effects of prey and weather on golden eagle reproduction. Journal of Animal Ecology 66, 350–362.*

Watson, A, Rae, S. and Payne, S. (2012). *Mirrored sequences of colonisation and abandonment by pairs of golden eagles Aquila chrysaetos. Ornis Fennica 89, 229–232.*

Watson, J. (1997). *The golden eagle. Poyser, London.*

Watson, J. and Dennis, R.H. (1992). *Nest site selection by golden eagles Aquila chrysaetos in Scotland. British Birds 85, 469–481.*

Watson, J., Rae, S.R. and Stillman, R. (1992). *Nesting density and breeding success of golden eagles (Aquila chrysaetos) in relation to food supply in Scotland. Journal of Animal Ecology 61, 543–550.*

Whitfield, D. P., Fielding, A. H., Gregory, M. J., Gordon, A. G., McLeod, D. R. and Haworth, P. F. (2007). *Complex effects of habitat loss on golden eagles Aquila chrysaetos. Ibis 149, 26–36.*

6. A DECLINE IN HOME-RANGE OCCUPANCY & BREEDING SUCCESS OF GOLDEN EAGLES IN THE DURNESS AREA OF NORTH-WEST SUTHERLAND

Stuart Rae, Adam Watson & Derek Spencer

SUMMARY

The occupancy of four golden eagle *Aquila chrysaetos* home ranges in the Durness area of north-west Sutherland was monitored in 22 years between 1958 and 2014, and the breeding success for 18 years. Home-range occupancy was 25% less than in 1958. The number of young reared per golden eagle pair and the number of pairs that reared young decreased between 1958 and 1985. There were more breeding failures after the mid-1970s. Disturbance of breeding eagles increased in later years. More pairs failed to rear young at undisturbed nests in later years. This decrease might have been linked to a decrease in food availability.

INTRODUCTION

Rae & Watson (Chapter 5) described home-range occupancy and breeding success of golden eagles in north-west Sutherland and compared data between two study periods, 1957–1960 (Brown & Watson 1964) and 1982–1985 (Watson *et al.* 1992). They found that eagle home-range occupancy dropped between study periods; fewer eagles laid eggs; and more eggs were broken in the earlier period, having probably been affected by DDT/Dieldrin, as elsewhere in the Highlands (Lockie & Ratcliffe 1964, Lockie *et al.* 1969). However, the breeding success was no greater in the later period, due to higher losses caused by disturbance.

The previous paper described data and results from only two time periods, with a gap of 22 years and ending in 1985. The present study used data from fewer home ranges for more years up to 2014, with the aim of finding whether there was any particular year or period with a change in home-range occupancy and breeding success. This study was of golden eagle pairs that bred in home ranges within 20km of the village of Durness in the far north-west of Sutherland. The study area had similar habitat and variety of potential prey and carrion to those in the larger previous study. It was closer to the coast, with less extensive high ground, although it still ranged from sea-level to over 900m.

METHODS

The previous study used data from 16 home ranges over two short four-year periods. The current study followed the same approach, using a longer run of data from fewer, four, contiguous home ranges within the same wider study area over 56 years between 1958 and 2014. Data for the additional years were collected by Bernard Hendy (1962–1981) and Derek Spencer (2011–2014). The number of home ranges monitored in each year varied, and the analyses followed the same sequence of reasoning as in the previous study (Rae & Watson–Chapter 5). We analysed data using correlation/ regression whenever suitable and supported with two-tailed t-tests for differences between periods when appropriate.

Tests were also made for any effect of disturbance of breeding attempts in relation to accessibility; i.e. the distances of nest sites from roads and climbing ability required to reach the nest sites. Distance from the nearest public road was measured in a straight line from a nest to the nearest point of a road or rough track open to the public (Watson & Dennis 1992). Climbing accessibility was assessed in a three-point scale: 1) easy access without a rope, 2) simple access with a rope, and 3) inaccessible without mountaineering rope techniques (Brown 1969, Watson & Dennis 1992).

RESULTS

HOME-RANGE OCCUPANCY

A maximum of four home ranges were occupied in any year. The year with fewest occupied ranges had two (1992), and in 1983 there were three as in 2011–2014. We used ANOVA to test for differences between four periods; 1958–69, 1970–79, 1981–85, 2011–14, (F = 21.04, df =3,19, P < 0.001), and occupancy in 2011–14 was significantly less than in all other periods (P < 0.01 for all interactions). In 2011–2014, two home ranges were occupied by only one pair for more than a three-year period. Therefore, we assessed one as abandoned (Whitfield *et al.* 2007), and amalgamated the home ranges to be used by only one pair. This came to a loss of 25% of the number of breeding golden eagle pairs in the area between 1958 and 2014, with the first decline in the 1980s.

BREEDING SUCCESS

This was calculated and analysed for only 18 years between 1958 and 1985 because there were too few data from the other years. We analysed breeding success of all pairs in the study area within two criteria: the total number of young reared per pair and the number of young reared per successful pair. Fewer young were reared per breeding pair in later years (r = -0.57, P = 0.01, df = 16, Figure 6.1) and fewer pairs reared broods of two (r = -0.57, P = 0.01, df = 16).

By Leslie Brown's criteria, the eyries on this Sutherland cliff are classed as unreachable without advanced climbing techniques. The authors agree.

This eyrie, the large column of sticks and heather stems (left of upper centre) is about two metres tall and built tight beneath a large overhang, well protected from the weather, although reachable with careful climbing or abseil from above. SR cleaned out a crack to fix a belay for abseil in the 1980s. On return with Derek Spencer in 2015 the crack was still clean and Derek used it once

again as an abseil point. It was surprising how little moss had grown around the crack over the years.

This nest was used in the year previous to when the photograph was taken, and the eagles have not re-touched the eyrie since. The top is still flattened as it had been by the chick before it fledged, and a bone from a prey item lies on the edge, perhaps the chick's last home meal. Long-unused eyries fade to grey piles of heather and sticks, but the ledges attract eagles for generations, and have probably been used for thousands of years since the first prey species colonised the area after the last ice-age.

This eyrie is on a small crag and although it is protected by an overhang, it can be viewed from the nearby hillside and accessed by rope. Eagles breeding there have failed in many years during the study.

This fledgling eagle is well sheltered by the overhang above its eyrie—it was raining when the photograph was taken and the photographer was not well sheltered, nor the camera, hence the soft grey hue. The flowers are honeysuckle, an unusual plant in the far north-western cliffs of Sutherland, and this is the only eyrie the authors know of where it grows so vigorously.

A range of sizes and colours of eggs of golden eagles. An average eagle egg from Scotland would be about 75 x 60 mm. All these eggs had failed to hatch and were collected once abandoned by the adults. The two on the top right are a clutch of two and both are small eggs, but the smallest is the one on the bottom left. In general, larger eggs offer the chicks that hatch from them a better chance of survival.

A clutch of two golden eagle eggs that were abandoned by the adults. Reasons for failure at this stage in the breeding cycle can be bad weather or human disturbance related. We cannot control the weather, but we can limit disturbance.

The remains of a golden eagle chick recovered from an eyrie by Derek Spencer. By watching the eyrie from a few kilometres away, and not causing any disturbance to the birds—they were seen to go in and out of the eyrie on several occasions—it became obvious that there was something wrong at the nest as the birds had stopped incubating or brooding and were both seen soaring over the adjacent hills and not approaching the nest to tend eggs or young. So, several days later Derek abseiled down to the eyrie to investigate. He found this headless and limbless chick, which had been only a few days old when it died. There might not have been enough food for the chick.

A healthy three week old chick sits up in its eyrie. The whitewash of droppings sprayed across the back wall is a good indicator that there is a chick in an eyrie when spied from a distance. So, even if a chick lies low or sleeps in the eyrie, if one sees this whitewash, then it is worth waiting for a few hours, watching through a telescope from a safe long distance. Any chicks will eventually lift their head up and so be confirmed and counted. Or if lucky an adult will bring in food and their heads will lift straight away.

HUMAN DISTURBANCES

Only one clutch of eggs was known to have been taken, in 1978. Hence, theft of eggs was not considered a significant cause of breeding failure in the Durness area between 1958 and 2014.

The category 'all disturbances' included eggs robbed and any other disturbances that caused breeding eagles to fail to raise chicks, such as nests burnt and eggs abandoned. These incidents could have been intentional or unintentional, and caused by unknown persons on the land. There were too few records of each type of disturbance to analyse individual effects. Therefore, because the outcome was the same for all effects, we assessed all disturbances together.

The number of breeding attempts which failed due to disturbance in the later years rose significantly (r = 0.63, n = 18, P = 0.005, transformed data: log10 n+1, Figure 6.2). From figure 6.2, there were clearly more breeding failures after the mid-1970s than before and this difference was significant. Breeding failure due to disturbance after 1975 stood higher than before (t = 4.44, P < 0.01, df = 8, mean = 0 and 22.2, ± SE 5.01).

There was no difference between the distances of disturbed and undisturbed nest sites from the nearest public road (ANOVA, F = 1, P = 0.33, df = 1,16, n = 6, 12). The mean distance of disturbed nest sites was 2.78 km ± SE 0.59, range 1.68–5.07 km, and that of undisturbed nest sites was 3.94 km ± SE 0.75, range 0.54–8.67 km.

However, all nest sites known to have been disturbed at least once were within easy approach, i.e. within easy climbing access to the nest or within easy close approach to disturb breeding birds. All subsequent analyses of breeding success were done on data from breeding attempts uninfluenced by human disturbances.

BREEDING SUCCESS BY UNDISTURBED PAIRS

Between 1958 and 1985, only four nesting attempts produced two young eagles, all in the earlier years of the study (1958, 1960, 1968 and 1973). Three of these broods were in one home range, the two adjacent home ranges later amalgamated, and one in another. Any change in the proportion of pairs that did not lay eggs over the years could not be tested as there were too few records of whether eggs were laid or not. We found no broken eggs during the study, although there might have been some unrecorded because we did not access all nests to determine this for certain.

We compared the proportions of undisturbed breeding attempts that failed due to unknown natural causes pre and post 1975. More pairs failed in the years after 1975 (t-test: t = 4.65, df = 8, P = 0.01, mean = 0 and 24.9 ± SE 5.33, Figure 6.3).

DISCUSSION

Home-range occupancy by golden eagles at four long-established territories in the Durness area of north-west Scotland declined between 1958 and 2014. It is particularly poignant that one of the home ranges where eagles had been more productive, rearing broods of two chicks, is the one which was abandoned, or amalgamated with an adjacent range. In effect, a productive pair of golden eagles was lost from the national population.

The eagles' breeding success decreased after the mid 1970s, when a higher proportion of breeding attempts failed due to human disturbance and other unknown causes compared with prior years. Although this study used small annual samples, the findings agree with those of the previous study of larger annual samples over the wider study area (Rae & Watson–Chapter 5). The population and breeding success in north-west Sutherland had declined.

There is probably a baseline, normal, failure rate over the years, whereby some eagles in an area do not rear young for non-human-induced causes (Watson 1997). In the present study, the decline in breeding success was probably caused at least partially by disturbance of breeding eagles, where easy access to nest sites was a factor. This complements the same effect found by Brown (1969) and Watson & Dennis (1992). Brown surmised that people such as egg-collectors would probably rob nests nearest to roads or tracks, while shepherds or gamekeepers might interfere with easy sites but leave inaccessible nests alone. The increase in disturbance might have been a result of greater ease of travel by more people in later years when

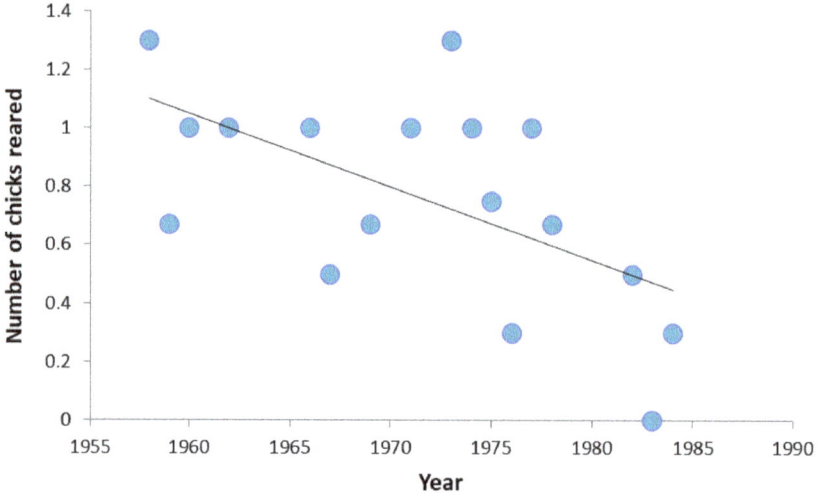

Figure 6.1. The numbers of young reared per breeding pair of golden eagles in the Durness area of Sutherland in 1958–85 (18 years).

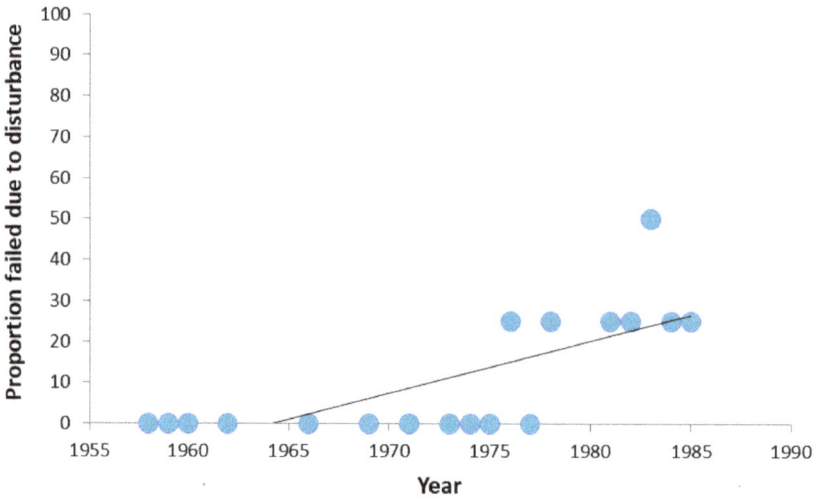

Figure 6.2. The annual proportions of breeding attempts by golden eagle pairs that failed due to human disturbance in the Durness area of Sutherland between 1958 and 1985.

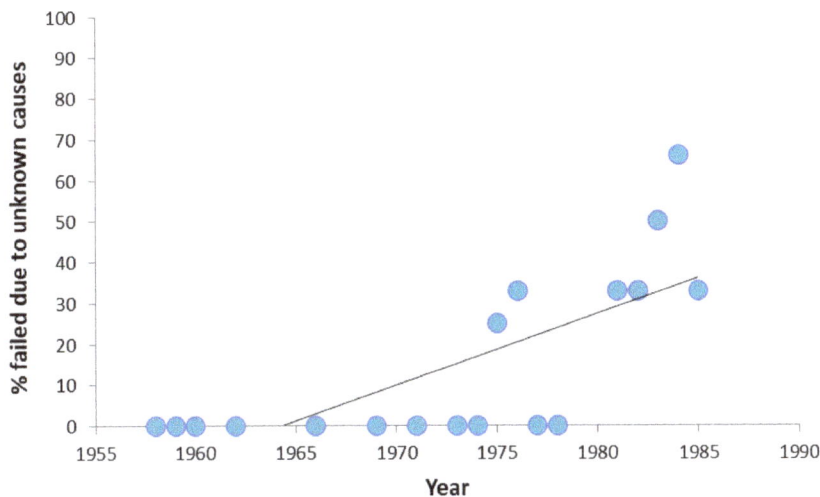

Figure 6.3. The proportions of undisturbed breeding attempts by golden eagles that probably failed due to natural reasons between 1958 and 1985.

tourism, hillwalking and bird-watching all became more popular in the Highlands during the period of the study. Numbers of people participating in these activities in the study area were not counted. However, there were sharp upturns of growth indices in mountain recreation in the mid 1970s (Neville *et al.* 2006) which coincided with the period when disturbance of breeding eagles increased.

There seems to have been an underlying unknown effect that appeared to cause the eagles to rear fewer young, even at undisturbed nests. Food supply might have had such an effect, because fewer broods of two chicks fledged in the later years. Although we did not test this, there was an apparent decline in the amount of available food between 1957–60 and 1982–85 (Rae & Watson–Chapter 7). Further analyses of breeding success and home-range occupancy, with comparisons of food availability over a long term of years, will be assessed in further study.

REFERENCES

Brown, L. H. (1969). *Status and breeding success of golden eagles in north-west Sutherland in 1967. British Birds 62, 345–363.*
Brown, L.H. and Watson, A. (1964). *The golden eagle in relation to its food supply. Ibis 106, 78–100.*

Lockie, J. D. and Ratcliffe, D. A. (1964). Insecticides and Scottish golden eagles. British Birds 57, 89–102.

Lockie, J. D., Ratcliffe, D. A. and Balharry, R. 1969. Breeding success and organo-chlorine residues in golden eagles in west Scotland. Journal of Applied Ecology 6, 381–389.

Neville, G., Duncan, K. and Mackay, A. (2006). Recreation. In: Shaw, P. & Thompson, D.B.A. (eds.). The nature of the Cairngorms: Diversity in a changing environment. The Stationery Office, 381–393.

Watson, J. 1997. The golden eagle. Poyser, London.

Watson, J. and Dennis, R.H. (1992). Nest site selection by golden eagles Aquila chrysaetos in Scotland. British Birds 85, 469–481.

An eagle's view down from the high tops of the Sutherland hills to the coast at the Kyle of Durness. The sea cliffs at the farthest headland are over ten kilometres away, mere minutes of flight for an eagle. Golden eagles can take ptarmigan on the rocky summits, seabirds from sea-cliffs or any other prey from the ground between.

7. FOOD AVAILABILITY FOR GOLDEN EAGLES IN NORTH-WEST SCOTLAND IN 1957-1960 & 1982-1985

Stuart Rae & Adam Watson

SUMMARY

We used data from two previous studies in 1957–60 and 1982–85 on the abundance of the main foods for golden eagles *Aquila chrysaetos* in north-west Sutherland. From these, it seems that there had probably been a decline, particularly of sheep carrion (65% less), and the number of sheep overwintered in the area had declined. Less prey was also found in the latter period. Although methods of counting potential food abundance differed between studies, the estimated amount per hectare in both studies was more than a calculated minimum requirement for a pair of eagles to raise a chick. However, these counts were done over the whole area and did not consider individual eagle home ranges, where there might have been differences in food availability and breeding success.

INTRODUCTION

The food supply of any animal population is fundamental to its demography, and Rae & Watson (Chapter 5) queried whether a reduction in food supply might have been a reason for a decline in the population and breeding success of golden eagles in their north-west Sutherland study area between, 1957–60 and 1982–85. Therefore, this study collated the amounts of foods found in the two study periods and discusses the possibilities of there being less food in the latter period.

The main foods eaten by golden eagles in the north-west Sutherland study area in 1957–60 were red deer *Cervus elaphus* carrion, sheep *Ovis aries* carrion, red grouse *Lagopus lagopus scoticus*, rock ptarmigan *Lagopus muta*, mountain hare *Lepus timidus* and rabbit *Oryctolagus cuniculus* (Brown & Watson 1964). In 1982–85, eagles ate a similar range of foods distributed almost evenly across these same six categories (Watson *et al.* 1987), with minor contributions from red fox *Vulpes vulpes* cubs, short-tailed field voles *Microtus agrestis*, fulmars *Fulmarus glacialis*, cuckoos *Cuculus canorus,* meadow pipits *Anthus pratensis,* and hooded crows *Corvus corone*. However, Watson *et al.* (1987) did not investigate whether there had been any changes in the availability of potential food items.

Here we discuss the possibility of any such effect.

The study area (Brown & Watson 1964, Watson *et al.* 1987) ranged from coast to hilltops above 900 m, so there was considerable variation in vegetation types and foods available. Further, there would have been variability in food in the various eagle home ranges. This was reflected in the foods eaten by eagles in each home range. In 1957–60, Watson found that birds at a home range bounded by sea cliffs took predominantly fulmars, and birds at five other sites near the coast took many fulmars, red grouse, and small numbers of shelduck *Tadorna tadorna*. Birds at a home range in the higher hill area took red grouse, rock ptarmigan, golden plover *Pluvialis apricaria*, and mountain hare.

Other foods eaten by eagles in the area, and not recorded during the two main studies, include one coastal pair that preyed on fulmars and barnacle geese *Branta leucopsis* in winter, and rabbits all year round, 1962–1981 (pers. comm. Bernard Hendy). Miscellaneous items taken in a later period, 2011–14, which would have also been available during the study periods and could have been taken then, included water vole *Arvicola terrestris*, common frog *Rana temporaria*, shag *Phalocrocorax aristotelis*, grey heron *Ardea cinerea*, and ring ouzel *Turdus torquatus* (pers. comm. Derek Spencer).

Although the availability of the main potential food for eagles was assessed in both study periods, the methods were not directly comparable. Therefore this paper aims to describe the possibility of, rather than test statistically, whether there was a change in potential food available between the two study periods, and whether this could have had an effect on the eagle population and breeding success.

METHODS

The study areas of Brown & Watson (1964) and Watson *et al.* (1987) were in the same region of north-west Scotland, although not in the exact same area. The latter area was larger and included all the ground covered by the former. It also held more eagles' home ranges.

The assessment methods were also dissimilar. Brown & Watson (1964, Tables 3 and 4) gave detailed data on food potential in their Area IV, i.e. the area they studied in north-west Sutherland. This comprised direct counts of the numbers of fully grown red grouse, ptarmigan, mountain hares, rabbits, dead sheep and dead red deer per 100 acres, seen by Watson with the help of trained dogs, on defined areas drawn on Ordnance Survey maps and later measured to the nearest 10 acres.

The method using dogs to quarter all the ground and scent or flush animals found was described briefly by Brown & Watson. A fuller description and tests of the

Charles Eric Palmar (Charlie) hangs over the top of an eagle cliff from a flimsily tied rope, while he takes a photograph with a heavy plate camera. (David Palmar, www.photoscot.co.uk)

Leslie Brown in the field. He always wore the kilt when walking and climbing, looking for eagles in the Highlands (Peter Steyn). He and Peter Steyn found this dying sheep on a hillside and as Peter recalls, 'He said, at the time that the sheep was dying of "pulpy kidney" whatever that may be and that it would make good carrion for golden eagles. On our way up to Argyll, we stayed with Charles Palmer in Glasgow who showed us his wonderful cine film of Golden Eagles.'

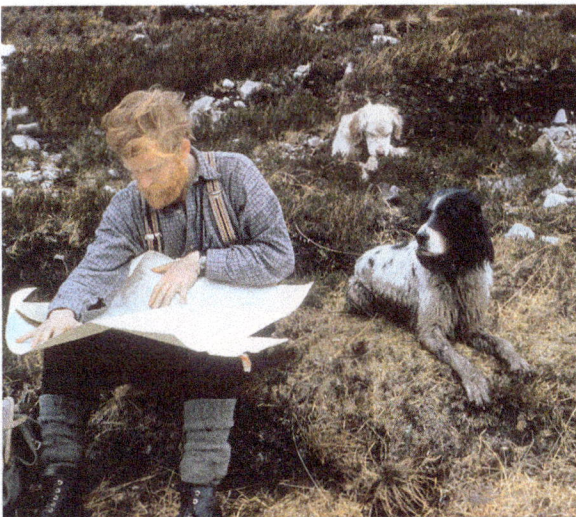

AW, checking a map while counting grouse in north-west Sutherland. He is accompanied by two English setters, Whisky alongside and Harra behind. (September 1959, AW senior)

Charlie examining a golden eagle eyrie and eggs. This was in exploratory days, 1951, with no safety equipment, but rolled up cuffs and open collar for climbing and a tweed jacket for warmth.
(David Palmar, www.photoscot. co.uk)

AW holds an eagle feather on a cliff-top where a golden eagle had plucked a fulmar, east of Clo Mor, the highest sea-cliff on mainland Scotland. These cliffs, which run along the northern coast of farthest north-west Sutherland, hold large colonies of seabirds with thousands of birds, including guillemots, razorbills, puffins and kittiwakes, but fulmars are the main eagle prey. In the 1950s and '60s the eagles that hunted there nested on a nearby inland cliff. They have not nested there for many years and the cliff has been occupied by common buzzards in recent years. (September 1958, AW senior)

validity of the method were published by Jenkins *et al.* (1963). Brown & Watson also presented (their Table 4) counts of the numbers of fully grown red grouse, ptarmigan, mountain hares, and rabbits seen per 100 miles by both authors during transect walks. They then estimated the mass of each food item in kg per 100 acres and kg per 2 miles. Watson *et al.* (1987) counted the same potential food items as Brown & Watson on pre-determined transect walk lines, without the aid of dogs. They assessed the items in kg per 100 km. Counts in both studies were done in spring, when any winter or early spring carrion would have been most obvious to surveyors, and prey species such as grouse and ptarmigan would have been dispersed within breeding territories. Both methods involved counts on glen bottoms, moors, and high tops. Both studies concentrated on the known main foods eaten by eagles in the study area. They did not include other potential food species found in the area such as golden plover or voles, nor species found only locally, such as fulmar.

Although, the survey methods of the two studies were not directly comparable, in this study we collated the abundance of potential foods recorded by the original studies (Brown & Watson, 1964, Watson *et al.* 1987), and then converted the differently formatted data into comparable units of measurement. From these data, we made approximate comparisons of food availability for golden eagles between the study periods. AW did a few counts in September 1958–60, outside the area of the earlier study, but within the 1982–85 study area, and we now give the results from counts where he still had field notes. These transects were all on the Parph, the moors west of the Kyle of Durness and west of the Durness to Rhiconich road. He did transect walks in September 1958 with an English setter dog ranging and with two setters and a pointer in September 1960. Then in September 1960 he and his father and three dogs walked transects over hilltops on the Parph.

Results

There was 53% less of the main potential foods available to golden eagles in 1982–1985 than in 1957–60 (Table 7.1). The largest component of this was 65% fewer sheep carrion, followed by 39% less prey. Only the amount of red deer carrion increased, by 13%. The amounts of all prey species decreased, with the main differences within them; 89% fewer mountain hares and 74% fewer ptarmigan. The amounts of rabbits detected decreased by 18% and that of red grouse by 7% (Table 7.2).

In the wider area, on the Parph, in 70 miles of transects in September 1958, AW saw 24 red grouse, seven mountain hares, and seven rabbits, the rabbits all being on small former fields. In September 1960 he saw only two red grouse, in 12 miles, far lower values for red grouse than those from transect counts on moors in Area IV of Brown & Watson (1964, Table 4), although the values for mountain hares and rabbits were larger.

On four miles of hilltops on the Parph, he saw 19–25 ptarmigan (minimum and maximum, with six birds perhaps being seen twice). This was a high density of ptarmigan, although a cautionary note is that the result was from one small hill on only one day in autumn when the ptarmigan would have been in post-breeding flocks. The main counts were all done in spring, when the grouse would have been distributed over the area in pairs on territories. It would therefore be unreliable and unjustifiable to compare these counts with the results of the much larger samples for tops in Area IV as documented by Brown & Watson. Therefore, these September counts are not comparable, but are the best available documentation for the time.

The annual food requirement of a pair of golden eagles, for them to rear a chick and to sustain an immature bird, is 321 kg within their home range (Brown & Watson 1964). They estimated that in north-west Sutherland there was approximately 20 times the required amount in 1957–60, at 49 kg/100 ha. In the present study we estimate that there were 10.5 kg/100 ha in the area in the 1980s, therefore by the same ratio as used by Brown & Watson, there was 4.7 times the required amount of food available in the 1980s. These figures do not include miscellaneous prey items, such as seabirds, corvids, fox cubs or other smaller birds and mammals known to have been eaten by eagles in the area. However, we suggest that there had indeed been a decline in the main food items between the study periods.

Table 7.1. The length of transects, mass of potential foods (prey, sheep and red deer carrion detected per 100 km of transect) and the density of breeding golden eagles in north-west Sutherland in the periods 1957–60 and 1982–85 (per Brown & Watson 1964, Watson *et al.* 1987).

	Length of Transects (km)	Prey kg/100km	Sheep Carrion kg/100km	Deer Carrion kg/100km	Total Food Items kg/100km	Golden Eagle Pairs / 1000km²
Brown & Watson 1964	664	26.5	202	30.8	258.8	19.0
Watson *et al* 1987	594	16.1	70.2	34.8	121.1	15.3

AW on the summit ridge of Foinaven with Harra, while counting ptarmigan. The hills Ben Stack and Quinag lie in the background. Such wonderful days on the hill more than balanced the many days of wind and rain that had to be worked through. (September 1958, AW senior)

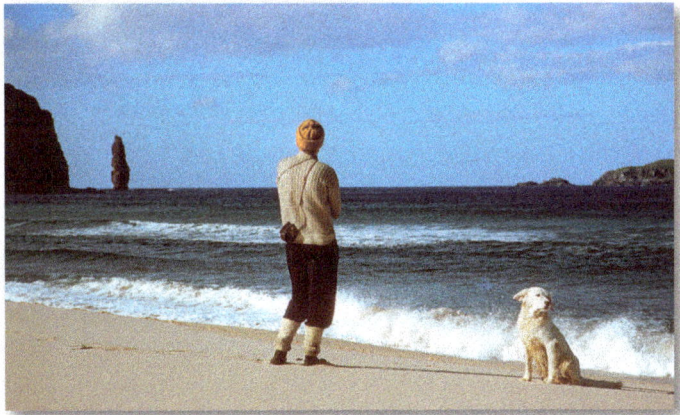

AW relaxes on Sandwood Bay beach after counting grouse with Harra. (September 1959, AW senior). The sea stack is Am Bauchaille and Tom Patey, who was a first ascensionist (1968), was a great friend of AW. Tom became a local doctor at Ullapool in the mid-late 1960s. AW explored the cliffs of north-west Scotland for nesting golden eagles, and Tom was a pioneering explorer of the cliffs for climbing. He died there on a climb in 1970. Most cliffs that eagles nest on are unsuitable for rock climbing as they are usually too vegetated, and climbers prefer clean, dry rock. Some eagle cliffs are suitable for climbing, and in other parts of the country some have been unused by nesting eagles since being developed as climbing crags. The eagle cliffs in the north-west that do hold good rock for climbing have to date probably been saved from climbing disturbance by the remoteness of the area and the distance of the cliffs from the nearest roads.

AW's wife Jenny at his old butcher's van, a Morris commercial J-type (1956), beside the quay for the passenger ferry across the Kyle of Durness. The ferryman's cottage can be seen in the background, on the far side of the kyle. When AW and Jenny slept in it on a hard frosty night, the metal in the roof fairly cracked, enough to waken them up. When the butcher had it, all the paintwork was a wishy-washy dark green, so AW covered it with shining black thick paint, but never got round to painting the roof.

The VW caravanette, which SR used as a mobile field base during the study years 1982-85. He and Jeff Watson used one each and were seldom in contact with one another in those days before mobile telephones. At any time, SR could be in Sutherland and Jeff in Ardnamurchan. The freedom to come and go as needed offered them the versatility needed to complete the colossal work load they undertook. They both covered three study areas, each with about 20 eagle home ranges. Rigorous standards of work and self-dependence were the main requirements, and the work got done. Today's standards of health and safety would be too prohibitive for people to work alone in remote terrain under such conditions.

John Scobie, Loch Stack (September 1958). John was head stalker on Reay Forest, which contained a large part of the north-west study area. AW received a great welcome when he first introduced himself and his purpose, including the eagle study and prey transects. He remembers getting tea, during which John's son Billy, stalker at Laxford Bridge, asked him to stay with him and his wife Cathy (free). It turned out that they had a high opinion of Derek Ratcliffe, who had been staying with them shortly before (paid by the Nature Conservancy), during his plant surveys of Sutherland. Derek also noted eagle eyries while on his surveys and passed the information onto AW. Hence these eagle studies in north-west Sutherland began.

Table 7.2. **The mass per length of transect (kg/100 km) of each main prey species of golden eagles detected in north-west Sutherland in 1957–60 (per Brown & Watson 1964) and 1982–85 (per Watson *et al*. 1987).**

	Brown & Watson 1964	Watson *et al* 1987
Red Grouse	10.5	9.8
Ptarmigan	6.6	1.7
Mountain Hare	4.4	0.5
Rabbit	5.0	4.1

DISCUSSION

In the 1980s, the amount of sheep carrion found in the study area had decreased to about one third of the amount found in the 1957–60 period. Although the two studies used different methods, it is questionable whether the difference between the potential food abundances would be so large if there had not been a true effect of material size. Sheep carrion would have been the most noticeable of all the food items because it consisted of white fleece, often scattered across metres of dark moorland vegetation (mostly heather *Calluna vulgaris*), it is long-lasting and does not require to be located by scent, as by a dog. Therefore, although the later figure might be less than what might have been found by dogs, most of the sheep carcasses would probably have been detected.

That there was a reduction in the amount of sheep carrion between the study periods is probably correct, because, over a similar period, there were fewer sheep on hill ground in the wider region in winter. In 1984 there was 64% of the number in 1964, within the local parish boundaries which covered the study areas (Watson *et al.* 1987)). Other factors might also have affected the amount of sheep carrion on the open hill, such as changes in sheep management or death rate at different stocking levels, but we did not study these.

There seem to have been fewer mountain hares, rock ptarmigan and red grouse, although these figures are much less reliable because dogs would detect more of these than a person walking a transect alone (Jenkins *et al.* 1963, Brown & Watson 1964). All these species' numbers are known to fluctuate (Jenkins *et al.* 1967, Watson & Hewson 1973, Watson *et al.* 1998). Therefore, their numbers might have been lower in the latter period.

The most noticeable point of the autumn counts is the lack of carrion. This contrasts with its abundance in the main survey, which fitted the expected abundance of carrion in late winter/spring (Watson *et al.* 1989). Brown & Watson (1964) found no evidence that food availability (prey or carrion) was a factor in the eagle breed-ing density (Brown & Watson 1964). However, Watson *et al.* (1989) and Watson *et al.* (1992) found that the density of breeding golden eagles was correlated with carrion availability. The dlanensity of golden eagles in the study area was lower in 1982–85 than in 1957–60 (Rae & Watson–Chapter 5). Therefore, if sheep carrion did decline, that factor could have had a negative effect on eagle density.

Few pairs of eagles raised more than one chick per annum in 1957–60 (Brown & Watson 1964). Intuitively, two chicks require more prey than one. Watson *et al.* (1992) found evidence to support this, and Rae & Watson (Chapter 5) found that fewer pairs of eagles reared young and fewer had broods of two in the study area in the 1980s than in the earlier period. Therefore, there could have been a real decline in the availability of prey. Also, the reduction in the amount of sheep carrion was so large, it should not be ignored that the amount of carrion available to eagles in late winter/early spring might also affect breeding success via the possibility of lay-ing fewer eggs or poorer-quality eggs. Or, as the area still contained 4.7 times the amount of food required by a pair of golden eagles to rear young, perhaps the pro-portionate quality of the food is more important to eagles than the total abundance.

Although this study was of the main food types eaten by eagles in the area, eagles in some home ranges ate wider diets. Some, for example, focused on fulmar or rabbit. An abundance of local prey can affect the local breeding success (Whitfield *et al.* 2007). A local abundance of prey might have given higher density and breeding success (Rae & Watson–Chapter 5, Rae *et al.*–Chapter 6) than if no eagles in the study area had access to such foods. Fulmar and rabbit are local in abundance in the study area and not all eagles had access to them. There might also have been local variations in the abundance of the main foods between eagle home ranges. However, the present study was of the main foods available to the population of eagles overall; it did not take into account any local variation in eagle diet between home ranges.

There does seem to have been a decrease in the amount of sheep carrion and pos-sibly a decline in the abundance of prey in the study area between 1957–60 and 1982–85, with the former possibly having a negative effect on the density of golden eagles between these periods (Rae & Watson–Chapter 5). Although the figures for the differences in prey abundance are uncertain, if the decline in breeding success (Rae & Watson–Chapter 5) were related to food supply, then it likely was affected by a decline in prey, as well as carrion. Firm conclusions will require further study.

ACKNOWLEDGEMENTS

Bernard Hendy and Derek Spencer provided extra notes on food eaten by golden eagles in the area.

REFERENCES

Brown, L.H. and Watson, A. (1964). *The golden eagle in relation to its food supply. Ibis 106, 78–100.*

Jenkins, D., Watson, A. and Miller, G. R. (1963). *Population studies on red grouse, Lagopus lagopus scoticus (Lath.) in north-east Scotland. The Journal of Animal Ecology, 317–376.*

Jenkins, D., Watson, A., & Miller, G. R. (1967). *Population fluctuations in the red grouse Lagopus lagopus scoticus. The Journal of Animal Ecology 36, 97-122.*

Watson, A. and Hewson, R. (1973). *Population densities of mountain hares (Lepus timidus) on western Scottish and Irish moors and on Scottish hills. Journal of Zoology 170, 151–159.*

Watson, A., Payne, A.G. and Rae, R. (1989). *Golden eagles Aquila chrysaetos: land use and food in northeast Scotland. Ibis 131, 336–348.*

Watson, A., Moss, R. and Rae, S. (1998). *Population dynamics of Scottish rock ptarmigan cycles. – Ecology 79, 1174-1192.*

Watson, J., Langslow, D.R. and Rae, S.R. (1987). *The impact of land-use changes on golden eagles in the Scottish Highlands. CSD Report no. 720. Nature Conservancy Council, Peterborough.*

Watson, J., Rae, S.R. and Stillman, R. (1992). *Nesting density and breeding success of golden eagles (Aquila chrysaetos) in relation to food supply in Scotland. The Journal of Animal Ecology 61, 543–550.*

Whitfield, D. P., Fielding, A. H., Gregory, M. J., Gordon, A. G., McLeod, D. R. and Haworth, P. F. (2007). *Complex effects of habitat loss on golden eagles Aquila chrysaetos. Ibis 149, 26–36.*

8. COMMENTS ON THE STATUS OF GOLDEN EAGLES IN SCOTLAND

Stuart Rae & Adam Watson

A large part of this volume is archival information on golden eagles *Aquila chrysaetos* in Scotland. We learn from history; hence the longer the run of data, the greater the history and the greater our knowledge. It is from a thorough understanding of how eagles fit in the landscape, and have fitted in the past, that we can anticipate how they will probably fit and survive in the future.

We deliberately use the term "landscape" for the environment that golden eagles live in, because that has connotations of a human perspective of the land. All the land, and air, that eagles live in has been altered by human use. That is the state of things, and eagles cannot be considered separately, not even in what many people regard as the wild Scottish Highlands. Yet eagles persist there, thrive there in some places and in some times, and the information in the previous chapters shows some aspects of how they do so. The historical high density of breeding golden eagles in our north-east Scotland study area might have been an encouraging example for a healthy population, but it was unnatural, and short-lived. It was associated with prehistoric massive deforestation and later high abundance of deer carrion, which resulted from management that encouraged high numbers of deer (Watson & Rae–Chapter 2). Any lesser amounts of deer carrion and prey, and increases of woodland would probably lead to a reduction of density and breeding success of the remaining pairs of golden eagles.

Eagles eat a variety of foods in Scotland, and the main prey are red grouse *Lagopus lagopus scoticus*, rock ptarmigan *Lagopus muta*, mountain hare *Lepus timidus* and rabbit *Oryctolagus cuniculus*, with carrion from dead sheep and red deer *Cervus elaphus* (Watson *et al.* 1993). They also eat many miscellaneous items and have different diets in several regions, depending upon the local environment and human land use (Brown & Watson 1964, Watson *et al.* 1992, Whitfield *et al.* 2006). There can even be differences in diets between adjacent home ranges if the local abundance of particular foods varies (Rae & Watson–Chapter 7). Differences in the availability of their food can affect the status of breeding eagles, and it is possible that some breeding pairs in Scotland rely predominantly upon a narrow diet. A decrease in the availability of any staple food might reduce those eagles' breeding capacity, and that can be assessed only if their diet and food availability have been recorded and any changes measured.

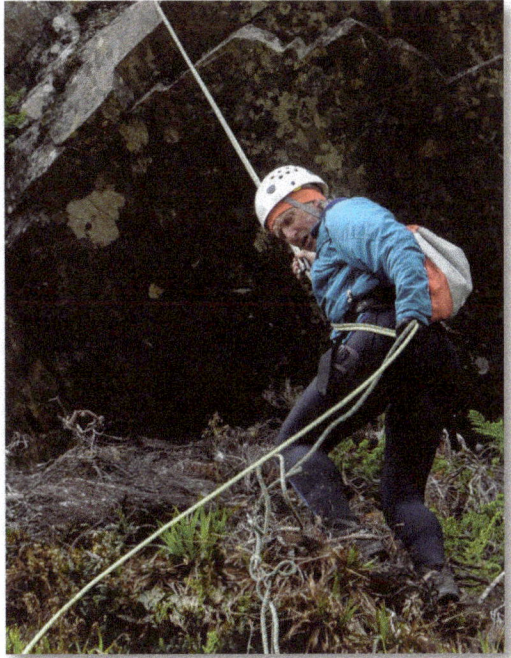

Derek Spencer abseils down from a golden eagle eyrie having checked the contents, in this case, a recently dead chick in the nest. Modern lightweight equipment makes climbing into eyries so much easier and safer than in the pioneering eagle days. The ropes are more pliable and longer, and helmets protect heads from rockfall, which is an ever-present risk on cliffs that have seldom if ever been climbed before.

Belaying methods are safer and much more adaptable and solid than in the days of Brown and Watson. They scrambled on wet vegetated ledges to reach some nests or roped into trickier sites (as did SR in the 1980s). Access then was only possible on the more moderate cliffs, while nowadays Derek has reached many more eyries to confirm whether they had been used or not and collect information on what the birds had been eating. However, even with modern equipment there are still many eyries out of sensible reach.

If there is enough food in an area, eagles can settle and rear young. That requires a nest site, which can be a cliff or a tree, even a hillside or small bank, as long as there is enough airspace for them to fly in and out. Shelter from weather is probably a requisite feature for a nest site, whereas safety from predators does not seem to be a recent priority for eagle eyries in Scotland. Eagles probably do not need to avoid the largest potential predators such as red fox *Vulpes vulpes*, although extinct larger predators such as wolf *Canis lupus* and lynx *Lynx lynx* would have been more of a threat. Today, human disturbance jeopardises eagles' use of many nest sites. Breeding attempts at easily accessed sites have failed and some sites are no longer used (Rae & Watson–Chapter 5, Rae *et al.*–Chapter 6). Any conservation plan for eagles should encompass all their requirements and take into account nest site types and disturbance.

The numbers of outdoor recreational visitors to the Highlands have increased during the span of the above studies; e.g. 4.2 million day visits for active or special interest pursuits in 1973 and 11.7 million in 1989 (CCS 1990). The numbers of hill-walking and climbing clubs affiliated with the Mountaineering Council of Scotland also increased, from about 20 in 1945 (Neville *et al.* 2006) to over 140 in 2016 (MCofS 2016). With such upsurges in numbers of people, it is expected that the potential for human disturbance of eagles would also have increased. In Finland, occupancy of golden eagle territories around tourist developments decreased during 1990–2004 and it was lower still around larger tourist developments (Kaisanlahti-Jokimaki *et al.* 2008). In Scotland, Whitfield *et al.* (2007a) did not find any

Two eaglets lie quietly on the ground below a tree eyrie, ready to be ringed and measured. It is better to lower chicks down from eagle eyries for ringing and measuring as they can be handled more securely and safely—for birds and personnel. This is not so easily done on cliffs due to the inconvenient rock structure, but is usually easily done from tree nests.

correlation between visitor numbers and abandonment of eagle nest sites, although that was a desk study using surrogate figures with no direct assessment and they suggested further analysis. In north-west Sutherland, the occurrence of disturbance was verified with fieldwork, and an increase of breeding failure due to disturbance coincided with an increase in mountain recreation (Rae & Watson–Chapter 5, Rae *et al.*–Chapter 6).

Also, the authors know eagle nest sites in other areas that have not been used for decades; they are of easy access, close to roads or tracks, or they are on cliffs used by rock and ice climbers. Birds still breed in some of those home ranges, but now use alternative nest sites farther from access points. Even there, they have probably been adversely affected by visitors. The birds would probably have been using the easily accessed sites for a reason which benefitted them. The farther or higher nest sites might not be as suitable for successful breeding. More study is required to document direct and indirect effects of recreational activities on the behaviour of golden eagles, including their use of nest-sites, to resolve any conflict between eagles and visitors.

In Scotland, in the eighteenth and nineteenth centuries, golden eagle populations were constrained due to persecution; in the twentieth century their numbers increased due to reduced persecution and enlightened attitudes towards eagles. After World War II, they suffered from the effects of organochlorines introduced to the ecosystem in pesticides.

SR closes a ring around a young eagle's leg. Eagles are strong birds, so the rings they are fitted with have to be strong too. Those used in Scotland are supplied by the British Trust for Ornithology, made from stainless steel and with a doubled-over fastening. It also takes strength to fasten them. (Simon Cherriman)

A recent innovation has been the addition of a uniquely alpha-numeric coded ring on the eaglet's other leg. These are fitted with rivets. The aim is that birds with such rings can be identified in the field by remote cameras set at feeding, roosting and nesting sites. The unfeathered heel becomes feathered after the eaglet fledges.

The following three photographs illustrate some measurements that are taken of the birds. By using these data it is possible to calculate the age of chicks by comparing them with those of chicks of known age. Then from that information it is simple to count back to record the likely hatching date and thus laying dates of the eggs.

Jenny Weston measures the maximum chord, a standard method of measuring a bird's wing. This bird's half-grown wing is 300 mm; that of a full-grown bird reaches about 600 mm. The wingspan of a golden eagle, the full breadth of both wings open, measures about 2 m.

The claw on the hind toe of an eagle's foot, the hallux, is the strongest and the longest. This bird's hallux claw is 40.8 mm long; an adult's would be about 50 mm.

The bill length is measured from the cere to the tip. This one is 33.6 mm long. Full-grown males have bills about 40 mm, females about 45 mm.

Two of the main negative effects on breeding golden eagles in north-west Sutherland have been non-laying in the 1950s and '60s and reduced breeding success in the 1980s (Rae & Watson–Chapter 5). The former was probably caused by contamination with organochlorine compounds, DDT or Dieldrin, the latter partially by human disturbance. Both causes were anthropogenic. A third unidentified factor also reduced breeding success, and that could have been the amount of food available to the eagles. If food availability were a problem, that too would have been a result of human land use, via land management practices such as sheep husbandry and deer culling. It should be possible to resolve any anthropogenic effects and, in the case of the organochlorines, it was, subsequent to the detection of the effect. That detection was possible only because breeding success was monitored before and after the introduction of the chemicals, thus illustrating the value of long-term study (Lockie & Ratcliffe 1964, Lockie *et al.* 1969).

The organochlorines that affected eagles were used in the sheep industry, and most eagle ground in the west Highlands is used for sheep ranching. Yet it has been long predicted that the high eagle stocks there could not exist without abundant sheep and associated abundant carrion (Lockie & Ratcliffe 1964, Brown & Watson 1964). There seemed to have been a large decrease in the amount of sheep carrion in the study area in north-west Sutherland between 1957–60 and 1982–85, which coincided with a drop in sheep numbers on the hills in the region (Watson *et al.* 1987). That decrease probably had a negative effect on golden eagle density. Studies of home-range occupancy and breeding success in that area are ongoing, with complementary assessment of sheep carrion or other potential food abundance. This is propitious, because there has subsequently been a decline in the number of sheep in Scotland since 1999, especially in hill areas after the change in agricultural payments in 2005. Currently there are between 0.1 and 1.0 breeding ewes per hectare in most of the

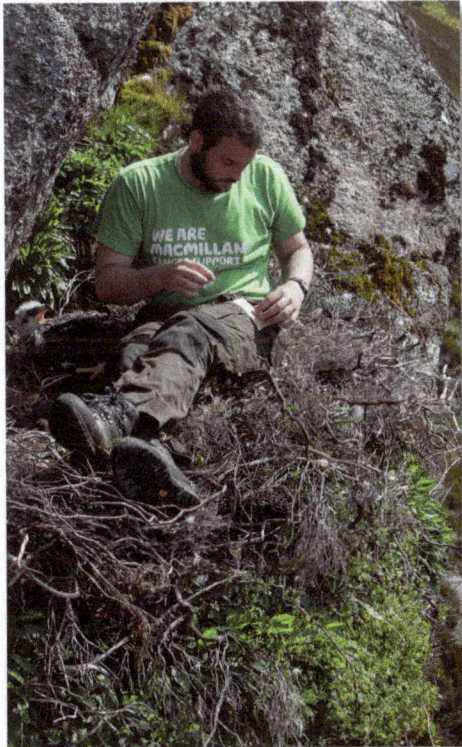

Ewan Weston sits on an easily accessible eyrie with an eaglet as he gathers data; there is ample room for two. Ewan is of the next generation of the eagle study dynasty to come from north-east Scotland and has now extended his study all over Scotland, just as AW and SR have done. The skills required nowadays include all the old standard fieldcrafts plus various forms of remote-locator-device fitting and DNA sampling. He does the climbing.

Highlands (RESAS 2015). As a result, it is expected that eagle numbers in much of the west Highlands will fall. This might not be immediate, because eagles are long-lived and they might continue to try to breed, probably with less success, or they might carry on as non-breeding pairs or single birds.

One of the main land-uses in the Highlands, deer stalking, is currently profitable, as is eco-tourism, where people come to see wild red deer. However, it is considered publicly and morally unacceptable to allow deer numbers to increase to such levels that they are subject to high mortality rates in winter (Edwards & Kenyon 2013). This has influenced many landowners to shoot more deer, which would reduce the amount of carrion for eagles over most years. However, there would still probably be occasional local abundances of carrion following winter storms, which would be more of a natural occurrence.

On some intermediate ground used for both deer and grouse, declines in grouse bags have led to less effort in moor management in east Scotland and subsequently to a natural spread of woodland onto moorland. The consequence has been reductions in the eagles' hunting grounds and in the densities of their main prey (Gillings *et al.* 2000). Also, numbers of grouse are lower on tick-infested moors, including part of the north-east study area. In the long run, there would probably be fewer red grouse and mountain hares in such places, and any adverse effects on eagles would probably develop slowly.

However, the main immediate impact on eagles in Scotland is persecution, which mostly occurs on moorland managed for grouse (Whitfield *et al.* 2003). This has a direct effect on the local birds, especially in the eastern Highlands where most persecution occurs, but it also potentially draws birds from farther afield, hence creating an ecological sink (Whitfield *et al.*

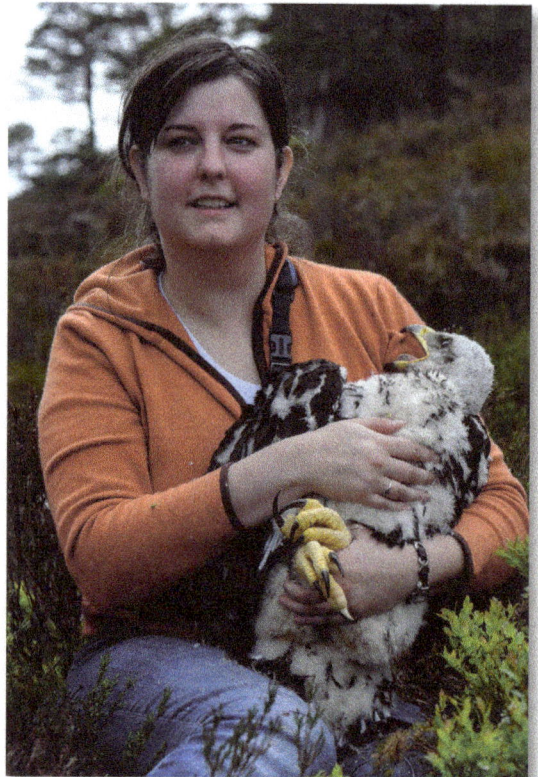

Jenny Weston, Ewan's wife, holds a large female eaglet ready for ringing and measuring. She does the measuring. Good teamwork makes eagle study quick and efficient, when the safety of the birds and personnel are important. Hence the gentle embrace to keep the bird from flapping, the firm grip on those talons and avoidance of that snapping bill.

Simon Cherriman, from Perth, Australia. He laughs as an eaglet has just evacuated its cloaca all down his front. Simon studies wedge-tailed eagles in Australia and has been across to Scotland to learn and share experiences on working with eagles. Scotland was one of the first places where eagles were studied and the standard of work is still at the fore of their worldwide study.

2004). As long as persecution continues, wherever a breeding eagle or a potential recruit to a breeding site is killed, a vacancy occurs in a breeding territory. Another eagle will be attracted to that site and if that too is killed, another will be attracted and killed, and so on.

Hence the impact of any persons killing one eagle after another on a local area is much greater than it might appear. The whole national population is affected. Also, when healthy birds are killed in those places, they are lost from potential recruitment into the wider population.

In a consultation response, AW wrote to the Scottish Government in 2011 expressing concerns about the prevalence of raptor persecution and the inadequate enforcement of the Wildlife and Countryside Act (Box A). Seven years later and 15 years after Whitfield *et al.* (2003), there has been little change. More than 200 years ago 'the eagle was persecuted relentlessly, and it is indeed remarkable that the species should have survived' (Gordon 1955). This problem still persists with no satisfactory answer. In Scotland, in the twenty-first century, golden eagles are still constrained due to persecution.

The current landscape where eagles live is human-induced and changing. Humans created large areas of open ground by clearing woodland, which enhanced the eagles' hunting grounds. And humans introduced the rabbit, so important as eagle prey. Much of the lower slopes where eagles now hunt would have been covered by woodland and held few eagles. It has been stated that the Highlands should be changed by preventing overgrazing by excessive numbers of sheep and deer, reducing careless, over-frequent burning, stopping mass intensive afforestation with densely planted conifers, and re-establishing natural woods (e.g. Dennis *et al.* 1984). This would create a more varied fauna, but would probably reduce eagle stocks. Tjernberg's (1985) study in Sweden, and personal observations by the authors of eagles' hunting behaviour in Scotland, indicate that eagles would be scarcer on densely regenerating natural woodland and scrub than on the present open moorland with some semi-open native woodland. This has happened elsewhere.

In the Italian Alps, declines of grazing cattle and sheep at high altitude have led to localised land abandonment for agriculture (Pedrini & Sergio 2001). In turn, this reduced the area of open pasture that affords good hunting ground for golden eagles, as scrub and trees encroached. Scotland's high eagle densities are as human-induced as the unusually high densities of red grouse, mountain hares, rabbits, red deer and sheep carrion upon which they largely depend. Many people wish to maintain present high eagle numbers, but this is antithetical to a diversified Highlands.

The issues of eagle numbers and land use are complex and varied across the country. It was expected that the status of golden eagles in Scotland in the 1980s would have been better than in the late 1950s and early 1960s, when there was a serious detrimental effect on their breeding success due to contamination with organochlorines. However, as has been shown (Rae & Watson–Chapter 5) that was not necessarily so. In north-west Sutherland the numbers of breeding pairs declined in the 1980s due to disturbance, and their breeding success also declined, as less food was available. Meanwhile, in the north-east, where organochlorines had little or no effect, the population decreased due to less carrion available for the birds; and clutch sizes were lower in the period after rabbit numbers declined in the north-east. These are examples of how numbers and breeding success of eagles in Scotland vary between biogeographical regions (Brown & Watson 1964, Watson *et al.* 1992, Eaton *et al.* 2007), and with the quality of each territory (Whitfield *et al.* 2007b). Judgement on how any local eagle population is faring should not be based on an

This is on old photograph of SR holding an eagle that had died after ingesting poison, but illegal persecution is still a disgrace, especially in parts of Scotland where moorland is managed for driven shooting of red grouse. (April 1988, Keith Brockie)

Not all eagle work involves handling birds. Most of it doesn't. The best way to watch them is from afar, because only undisturbed birds behave naturally. As in this case, where SR checks on a growing chick hidden out of reach on a cliff, and well away from human disturbance. Remote work in a remote place. (Derek Spencer)

Adam Ritchie and Vanessa Watson watch eagles on a distant cliff. Eagle watching in the Highlands involves many hours watching, watching, watching; waiting for the birds to show. SR has recently been studying eagles in another former study area of the 1980s in Ross-shire with Adam, Vanessa and Doug Mainland, comparing the breeding behaviour with that in the 1980s. They hope to test the results found in north-east Scotland and north-west Sutherland in a different area, and whether changes in the land use there also affect the golden eagle population.

The Eagle Hoose on Balmoral Estate, as this structure is colloquially called (hoose is Scots vernacular for house), where eaglets were taken from local nests in the late 1800s and kept in captivity until the early 1900s. The birds, kept for entertainment or prestige, had barely enough room to fly between perches. Collecting historical information is an integral part of current eagle study. John Robertson and AW with pointer Henry. (December 2012, Derek Pyper)

AW leans on a convenient outcrop of granite during a shower of rain in the Cairngorms. Some people joke and scoff at studying eagles in Scotland, where the weather is so often wet, windy and cold, and the birds are in such remote places. They ask how we do it, tolerate the weather or achieve any work, although clearly we do, as do other eagle devotees and other Highland naturalists. The Highlands have a viable population of eagles because of an abundance of food and nest sites, which are there due to the climate and geomorphology. They in turn are as they are because of where the land that is Scotland lies at the present time in the Earth's evolution. The ancient rocks have been weathered to form the hills and glens, lush plant growth is maintained by the rains brought in from the Atlantic Ocean by the westerly winds, and the abundance of prey and carrion is consequent to those conditions. So we tolerate the weather, and we are glad to be there because the Highland landscape is remarkable whether in sunshine or rain.

aggregate assessment of the national numbers but rather on a pair-by-pair basis.

The results of the studies presented in this volume derive from the long-term nature of the work. Golden eagles are long-lived birds and it is not possible to foresee what changes might occur in future. Only if long-term studies of eagles are continued into the future can changes in eagle demography be detected. There is today a body of volunteers who note data under the auspices of the Scottish Raptor Study Groups, and many of those members monitor a selection of eagle home ranges each year. However, few rigorously note data on food availability. Much of the information now being collected ignores other environmental factors which probably affect the golden eagles that they monitor. Also the national surveys of breeding golden eagles which were originally planned for every ten years have in actuality been done in 1982, 1992, 2003 & 2015. The timing of these surveys has been arbitrary and loose. It would be better if more reasoned and purposeful assessments of eagle numbers and breeding success were done in relation to conditions in the environment. For example, more rigorous data on food in the nest and its availability on the landscape could be recorded by observers who are in the field each year.

Data could be amassed from a set of biogeographic regions, and demography and food could be assessed in time scale and by region. Members of raptor study groups should aim to improve the suitability for purpose of the data they note. This process could be begun, e.g. in the six recently designated golden eagle Special Protection Areas (SNH 2010), where part of the required commitment to this international designation is site condition monitoring of the designated species (SNH n.d.). Al-

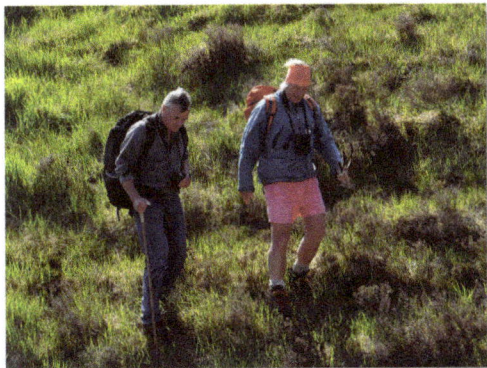

Des Thompson and Derek Spencer on a long walk out from an eagle eyrie. Studying eagles in Scotland demands long days, many of them. Long-term study is a bountiful system. As illustrated here in this book, it takes lifetimes and, we hope, generations of lifetimes to come.

though it is a small country, Scotland has sixteen biogeographic zones where golden eagles breed. Each zone has its own characteristics and constraints upon eagles, as do those in other parts of the bird's holarctic range. Study of golden eagles in a human-modified environment is complex, as is eagle conservation. Part of our research was based on historical information, part on long-term field study. It was possible to interpret links between golden eagles and human land use only with integral reasoning. The studies presented in this volume should help guide conservation of golden eagles in our study areas. Adaptation and further use of our methods and findings might also help enlighten conservation issues for golden eagles in other areas.

There are more than 100 years of information in this volume. If systematic data could continue to be noted, our results might be compared with data from 100 years or more into the future. We can learn from history and apply our knowledge to form a more responsible and appropriate future for golden eagles in Scotland. Scotland is the homeland, the fountain head, for eagle studies and eagle conservation worldwide, and the world is watching to see if we will develop and implement wise conservation practices.

REFERENCES

Brown, L.H. and Watson, A. (1964). The golden eagle in relation to its food supply. Ibis 106, 78–100.

CCS (1990). The mountain areas of Scotland: conservation and management. Countryside Commission for Scotland, Battleby.

Dennis, R.H., Ellis, P.M., Broad, R.A. and Langslow, D.R. (1984). The status of the golden eagle in Britain. British Birds 77, 592–607.

Eaton, M.A., Dillon, I.A., Stirling-Aird, P. and Whitfield, D.P. (2007). Status of the golden eagle Aquila chrysaetos in Britain in 2003. Bird Study 54, 212–220.

Edwards, Tom and Wendy Kenyon. (2013). SPICe Briefing: Wild deer in Scotland. Scottish Parliament, Edinburgh. http://www.scottish.parliament.uk/. [08 March

2016].

Gillings, S., R. J. Fuller, and D. E. Balmer. (2000). Breeding birds in scrub in the Scottish highlands: variation in community composition between scrub type and successional stage. Scottish Forestry 54, 73–85.

Gordon, S. (1955). The golden eagle. Collins, London.

Kaisanlahti-Jokimaki, M., Jokimaki, J., Huhta, E., Ukkola, M., Helle, P. and Ollila, T. (2008). Territory occupancy and breeding success of the golden eagle (Aquila chrysaetos) around tourist destinations in northern Finland. Ornis Fennica 85, 2–12.

Lockie, J. D., and Ratcliffe, D. A. (1964). Insecticides and Scottish golden eagles. British Birds 57, 89–102.

Lockie, J. D., Ratcliffe, D. A. and Balharry, R. (1969). Breeding success and organo-chlorine residues in golden eagles in west Scotland. Journal of Applied Ecology 6, 381–389.

MCofS (2016). Mountaineering Council of Scotland, clubs home page: http://www.mcofs.org.uk/clubs-home.asp/. [08 March 2016].

Neville, G., Duncan, K. and Mackay, A. (2006). Recreation. In: Shaw, P. & Thompson, D.B.A. (eds.). The nature of the Cairngorms: Diversity in a changing environment. The Stationery Office, 381–393.

Pedrini, P. & Sergio, F. (2001). Golden eagle Aquila chrysaetos density and productivity in relation to land abandonment and forest expansion in the Alps. Bird Study 48, 194–199.

RESAS (2015). Economic report on Scottish agriculture 2015 edition. Scottish Government, Edinburgh.

SNH (Scottish Natural Heritage) (2010). Classification of six golden eagle Special Protection Areas (SPAs). Available from: http://www.snh.gov.uk/protecting-scotlands-nature/protected-areas/international-designations/spa/classification/. [04 February 2016].

SNH (Scottish Natural Heritage) (n.d.). Site condition monitoring. Available from: http://www.snh.gov.uk/protecting-scotlands-nature/protected-areas/site-condition-monitoring/. [04 February 2016].

Tjernberg, M. (1985). Spacing of Golden Eagle Aquila chrysaetos nests in relation to nest site and food availability. Ibis 127, 250–255.

Watson, J., Langslow, D.R. and Rae, S.R. (1987). The impact of land-use changes on golden eagles in the Scottish Highlands. CSD Report no. 720. Nature Conservancy Council, Peterborough.

Watson, J., Rae, S.R. and Stillman, R. (1992). Nesting density and breeding success of golden eagles (Aquila chrysaetos) in relation to food supply in Scotland. The Journal of Animal Ecology 61, 543–550.

Watson, J., Leitch, A. F., & Rae, S. R. (1993). The diet of golden eagles Aquila chrysaetos in Scotland. Ibis 135, 387–393.

Whitfield, D.P., McLeod, D.R.A., Watson, J., Fielding, A.H. and Haworth, P.F. (2003). The association of grouse moor in Scotland with the illegal use of poisons to control predators. Biological Conservation 114, 157–163.

Whitfield, D.P., Fielding, A.H., McLeod, D.R.A. and Haworth, P.F. (2004). The effects of persecution on age of breeding and change in eagle distribution territory occupation in golden eagles in Scotland. *Biological Conservation* 118, 249–259.

Whitfield, D. P., Fielding, A. H., McLeod, D. R., Haworth, P. F. and Watson, J. (2006). A conservation framework for the golden eagle in Scotland: refining condition targets and assessment of constraint influences. *Biological Conservation* 130(4), 465–480.

Whitfield, D. P., Fielding, A. H., McLeod, D. R., Morton, K., Stirling-Aird, P. and Eaton, M. A. (2007a). Factors constraining the distribution of golden eagles *Aquila chrysaetos* in Scotland. *Bird Study* 54, 199–211.

Whitfield, D. P., Fielding, A. H., Gregory, M. J., Gordon, A. G., McLeod, D. R., and Haworth, P. F. (2007b). Complex effects of habitat loss on golden eagles *Aquila chrysaetos*. *Ibis* 149, 26–36.

At the end of the day, our reward for all the years of study is the satisfaction of seeing healthy eagle chicks reared, the national population maintained, and the knowledge that our work has helped and will continue to help the understanding of how golden eagles fit in the Highland landscape.

Box A. Letter to the Scottish Government: Wildlife and Countryside Act, consultation (abridged)

30 May 2011

'I request members of the Rural Affairs & Environment Committee to take the decisive action that is long overdue to end the widespread and increasing illegal persecution of protected Scottish raptors. Leading MSPs and Scottish Ministers in the previous Labour/Lib Dem coalition and the present SNP administration have condemned this unlawful activity, calling it a national disgrace. However, it continues unabated, involving poisoning as well as shooting and other illegal activities.'

'An example was in 1989 when I was author of a scientific paper that reviewed data on golden eagles from 1944 to 1981. It showed the adverse effects of eagle persecution on grouse moors, in contrast to deer forests where the birds were generally protected or ignored by resident deerstalkers. Also it showed how a decline of such persecution during the Second World War, when many grouse-moor keepers were in the armed services, led to a notable increase of resident eagle pairs on grouse moors. There followed a collapse over several years after the keepers returned. I was a founder member of the North East Raptor Study Group, the first of its kind in Britain. Many voluntary and professional observers now study Scottish raptors. As a result, sound information on their abundance and distribution has never been better. The data unquestionably show large recent declines in the abundance and distribution of golden eagles, peregrine falcons, hen harriers, and goshawks in northeast Scotland, primarily on grouse moors. The decline of hen harriers is catastrophic.'

'The buzzard shows what could be enjoyed by the public if persecution ended. Though formerly killed widely, it has returned in strength, especially on lowland, and many people appreciate it. Sparrowhawks and peregrines have also increased on lowland, and give pleasure to many. More enjoyment would come if eagles returned to eastern moors and woods. These examples illustrate the real, though hidden, costs to the public of raptor-killing.'

'Most conflicts can be resolved if both sides meet and compromise, and there have been meetings on raptor-grouse conflicts. When one side practises illegal acts, however, it is difficult to see whether the public would approve any compromise. Other suggestions to end persecution include legislation with mandatory jail-sentences, removal of gun-licences, game-shooting permitted by a licence that could be removed, and cross-compliance, e.g. denying state grants and exemptions of inheritance tax to owners of estates where the law on raptors has been broken. To conclude, grouse-shooting could continue without illegal persecution of raptors, but its future seems insecure if persecution continues unabated.'

Adam Watson

APPENDIX I

A CHRONOLOGICAL OUTLINE OF SOME IMPORTANT PUBLICATIONS ON GOLDEN EAGLES IN SCOTLAND

1855   MacGillivray, W. The Natural History of Deeside and Braemar. Printed for private circulation, Bradbury & Evans, Printers Extraordinary to the Queen, London.

First account of golden eagles in the north-east study area referred to in our book, describing the birds as very scarce due to destruction by shepherds and gamekeepers.

1907   Gordon, S. Birds of the Loch and Mountain. Cassell and Company, Limited, London.

Seton Gordon's first book, describing wildlife in the Highlands, with an opening chapter on the golden eagle.

1909   MacPherson, H. B. The Home-Life of a Golden Eagle. Witherby & Co., London.

The first monograph on the golden eagle, with notes and exceptional photographs of the birds and their behaviour at the nest.

1927   Gordon, S. Days with the Golden Eagle, Williams & Norgate Ltd., London.

Seton Gordon's first book on the golden eagle, describing the bird in Scotland and illustrated with his own photographs.

1955   Gordon, S. The Golden Eagle. King of Birds. Collins, London.

An absorbing account of the golden eagle, based on personal observations in Scotland with information and photographs from others in Scotland and around the world.

1957 Sandeman, P. W. The breeding success of golden eagles in the southern Grampians. Scottish Naturalist 69, 148–152.

One of a pair of papers presenting the first scientific accounts of the breeding biology of golden eagles in Scotland.

1957 Watson, A. The breeding success of golden eagles in the north-east Highlands. Scottish Naturalist 69, 153–169.

The second, and fuller, scientific account on the breeding biology of the golden eagle in Scotland

1964   Lockie, J. D., and Ratcliffe, D. A. Insecticides and Scottish golden eagles. British Birds 57, 89–102.

A ground-breaking paper on how chlorinated insecticides adversely affected breeding golden eagles, which ultimately contributed to a ban on these chemicals' use.

1964   Brown, L.H. and Watson, A. The golden eagle in relation to its food supply. Ibis 106, 78–100.

A thorough study of golden eagles' food, breeding density and success in four areas, which pioneered a scientifically strategic approach to understanding the ecology and conservation management of eagles in Europe.

1969   Brown, L. H. Status and breeding success of Golden eagles in northwest Sutherland in 1967. British Birds 62, 345-363.

An important stepping stone to the study of eagles in north-west Sutherland in our book. An early classic paper on the 'character of territory related to success'.

1974   Nethersole-Thompson, D. & Watson, A. The Cairngorms: their natural history and scenery. Collins, London.

A comprehensive description of the geomorphology, wildlife, land use, conservation and the future of the Cairngorms, which at the time was one of the least disturbed areas with eagles in Scotland. Most of the problems described have exacerbated, not diminished since its publication.

1976 Brown, L. British Birds of Prey. Collins, London.

A classic New Naturalist book. In this account of past and present status of raptors in Britain, Brown estimated that the population of golden eagles in Scotland would once have been about 500 pairs, and many nest sites would have been traditional. The current population has just attained this number.

1984 Dennis, R.H., Ellis, P.M., Broad, R.A. and Langslow, D.R. The status of the golden eagle in Britain. British Birds 77, 592–607.

A report on the findings from the first attempt, in 1982, to survey as much of the Scottish Highlands for golden eagles as possible. The total came to 424 pairs of eagles. In the last survey in 2015, 508 pairs were recorded.

1985 Marquiss, M., Ratcliffe, D.A. and Roxburgh, R. The numbers, breeding success and diet of golden eagles in southern Scotland in relation to changes in land use. Biological Conservation 34, 121-140.

A timely assessment of the recolonisation of the Southern Uplands of Scotland by golden eagles in the 1940s, complete with an accurate prediction of their subsequent demise.

1986 Watson, A. and Rothery, P. Regularity in spacing of Golden Eagle Aquila chrysaetos nests used within years in northeast Scotland. Ibis 128, 406-408.

A key paper outlining the regularity of territory occupation by golden eagles.

1987 Watson, J., Langslow, D.R. and Rae, S.R. The impact of land-use changes on golden eagles in the Scottish Highlands. CSD Report no. 720. Nature Conservancy Council, Peterborough.

A study of food, breeding density and success of golden eagles in six areas differing ecologically. This study followed the example set by Brown and Watson (1964), and covered 137 pairs, a third of the known Scottish population. Most of the identified land use effects are still applicable.

1989 Watson, A., Payne, A.G. and Rae, R. Golden eagles Aquila chrysaetos: land use and food in northeast Scotland. Ibis 131, 336-348.

A description of how numbers of breeding eagles declined when there was less carrion available after more red deer were shot, and the eagle numbers on moorland managed for shooting of red grouse shooting declined after 1946 due to persecution. Eagles bred well on land managed for deer, but poorly on areas of grouse moor, due to persecution.

1990 Ratcliffe, D. Bird life of Mountain and Upland. University Press, Cambridge.

An overview of the bird life of the hills in Britain, describing their ecology in context of environmental factors and the history of human land-use. Several pages are given to an important overview of the golden eagle.

1991 Newton, I. and Galbraith, E.A. Organochlorines and mercury in the eggs of golden eagles Aquila chrysaetos from Scotland. Ibis 133, 115-120.

Analyses of unhatched golden eagle eggs from 1963-86 showed that derivatives of organochlorines had declined but were still present, PCBs had increased mainly in the western districts, and in 1981-86 mercury residues were also detected at low levels in western but not eastern districts. Monitoring of chemical residues in eagles continues.

1992 Watson, J., Rae, S.R. and Stillman, R. Nesting density and breeding success of golden eagles (*Aquila chrysaetos*) in relation to food supply in Scotland. Journal of Animal Ecology 61, 543–550.

Golden eagle nesting density was found to be correlated with the different amounts of carrion in nine study areas, and breeding success was correlated with the amount of prey (grouse, hares and rabbits).

1993 Watson, J., Leitch, A. F., & Rae, S. R. The diet of golden eagles *Aquila chrysaetos* in Scotland. Ibis 135, 387–393.

Summer and winter diets of golden eagles were described for nine ecological regions in the Highlands and Islands of Scotland, based on 1793 items found in pellets. Six distinct regional diets were recognised, with different proportions of deer and sheep carrion, lagomorphs, grouse, seabirds and miscellaneous items.

1997 Watson, J. The golden Eagle. Poyser, London.

A rich assembly of scientific findings on the golden eagle, drawing on Scottish and wider global research. The book was updated in 2010, published after the untimely death of the author.

1998 Halley, D.J. Golden and White-tailed Eagles in Scotland and Norway. Co-existence, competition and environmental degradation. British Birds 91, 171-179.

An important paper which considers the potential for competition between the two species, drawing on experience of the species' co-existence in Norway.

1999 Grant, J.R. & McGrady, M.J. Dispersal of golden eagles *Aquila chrysaetos* in Scotland. Ringing & Migration 19, 169-174.

A report on the first use of radio telemetry to study the dispersal of young golden eagles in Scotland.

2003 Whitfield, D.P., McLeod, D.R.A., Watson, J., Fielding, A.H. and Haworth, P.F. The association of grouse moor in Scotland with the illegal use of poisons to control predators. Biological Conservation 114, 157–163.

This study confirmed, from analysis of data on poisoning incidents reported in Scotland between 1981-2000, that illegal poisoning in the uplands occurs disproportionately on land managed as grouse moor.

2005 Walker, D., McGrady, M., McCluskie, A., Madders, M. and McLeod D.R.A. Resident golden eagle ranging behaviour before and after construction of a windfarm in Argyll. Scottish Birds 25, 24–40.

The first study of the ranging behaviour of a pair golden eagle before and after the construction of a windfarm in the Scottish Highlands. The birds changed their flight activity away from the area occupied by the windfarm.

2007 Ratcliffe, D. Galloway and the Borders. Collins New Naturalist Library, Collins, London.

An evocative account of the plight of the tiny remnant population in south Scotland, which has suffered from illegal persecution and afforestation. Pages 238-244 are devoted to eagles, and in addition to a very good historical account of occupation, he has a Table (3) providing details of breeding success of the four territories in SW Scotland for 1945-2005.

2008 Whitfield, D.P., Fielding, A.H., McLeod, D.R.A. and Haworth, P. A. A conservation framework for the golden eagle: implications for the conservation and management of golden eagles in Scotland. Scottish Natural Heritage Commissioned Report No. 193 (ROAME No. F05AC306).

This report gave an analysis of the then current conservation status of the golden eagle in Scotland, identifying the main constraints as illegal persecution and low food availability in consequence of heavy grazing by sheep or red deer. A list of detailed proposals included the importance of maintaining suitable conditions for eagles beyond protected areas and research on other constraining factors.

2010 Bourke, B.P., Frantz, A.C., Lavers, C.P., Davison, A., Dawson, D.A. and Burke, T.A. Genetic signatures of population change in the British golden eagle (*Aquila chrysaetos*). Conservation Genetics 11, 1837-46.

The first estimate of genetic variety of the golden eagle population in Scotland, UK and Ireland using contemporary blood and feather samples and historical records from museum collections.

2012 Watson, A, Rae, S. and Payne, S. Mirrored sequences of colonisation and abandonment by pairs of golden eagles *Aquila chrysaetos*. Ornis Fennica 89, 229–232.

A description of the probable hierarchy in territory of golden eagle territory quality, from records of their historical sequence of use and disuse.

2012 Rae, S. Eagle Days. Langford Press, Peterborough.

A portrayal of studying golden eagle behaviour and ecology in Scotland, using diary notes and photographs to complement the more scientific description in Jeff Watson's monograph. The two authors studied eagles together.

2013 Watson, A. Golden Eagle colonisation of grouse moors in north-east Scotland during the Second World War. Scottish Birds 33, 31-33.

Documentation of twenty-two pairs of golden eagles colonising moorland areas managed for shooting red grouse when gamekeepers were away in the armed services and there was less persecution. Then the subsequent disappearance of 20 of those pairs, associated with the return of the gamekeepers and persecution.

2013 Weston, E.D., Whitfield, D.P., Travis, J.M. and Lambin X. When do young birds disperse? Tests from studies of golden eagles in Scotland. BMC Ecology 13, 13-42.

Calculations of the time taken by young golden eagles to disperse from their natal sites. The first description of natal dispersion of golden eagles in Scotland using GPS satellite tracking.

2015 Wikinews. Scottish gamekeeper jailed for bird crime in national first. https://en.wikinews.org/wiki/Scottish_gamekeeper_jailed_for_bird_crime_ in_national_first. Accessed 10 February 2017.

A press report on the first incidence of a gamekeeper imprisoned for crimes against birds of prey. In this case it was a goshawk that was killed. It took 160 years between MacGillivray recognising the problem and the first custodial sentence for the killing of a bird of prey in Scotland. Yet they continue to be illegally persecuted. It has for too long been obvious in the public interest for the Scottish Government to introduce such penalties, not only on estate staff, but on landowners of ground where persecution has been proved.

Malcolm Appleby MBE crafted this brooch with a golden eagle design for Adam, in thanks for their long friendship and shared appreciation of the bird.

Malcolm is known primarily as an engraver and is considered to be one of the most original and highly skilled craftsmen working in Britain today. His career spans over 50 years, promoting engraving through his own work and providing opportunities for young engravers and silversmiths.

There has been an obvious increasing number of publications on Scottish golden eagles over the years, reaching the heights since 2000. Hence, we understand far more about our golden eagles, and all the people who have written the above and many other published works on Scottish eagles have helped others understand golden eagles, all eagles, and other birds around the world. The output has been prodigious, as befits such a magnificent bird, and the findings reveal how eagles have at times and places struggled to fit with human land use. Throughout this history, there have been ups and downs, but there has been one recurrent solvable problem. Eagles in Scotland have been persecuted by man. After more than a century of awareness, the state of affairs for golden eagles in Scotland is still lamentable.

"My first close-up encounter with a golden eagle on Deeside was at Brathens Research Station in the 1970s. I believe the tethered bird belonged to Douglas Weir and was being looked after while Douglas was away. The sketches that I produced from that day were unremarkable; however to be able to observe at close quarters this huge bird with its wings outstretched sunning itself, its sharp piercing eye carefully watching its watcher is something that I will never forget.

I stumbled on eagle roosts while brown trout fishing and glimpsed an eagle slip out of the lone Scots Pine tree, almost unnoticed, a surprise for such a large bird. And the pleasure of adding treasured eagles' feathers to my collection contributes to my fond memories of Deeside. I also saw my first ravens, with eagles, on Deeside while out with Adam one January day just before a trip to London. I always cherished these excursions to the hills with Adam as he is an eagle-like spirit as much in tune with his landscape as the environment he studies. Ravens are more common now and thankfully for me an almost everyday experience at my new home in Perthshire. Eagles are of course much rarer sightings, but to experience eagles playing with ravens high in the clouds over the Cairngorms is a special privilege. I hope the eagle brooch conveys my first close encounter with this magnificent bird.

I'm passionate about the amount of energy and power that can be got out of the ancient craft of engraving. Engraving is central to my design and art; it is from engraving that my other skills have evolved. I enjoy harnessing other craftsmen's skills to extend my own creative theories of form-related goldsmithing techniques."

Malcolm Appleby August 2017

Fieldwork, like all trades, requires a good eye, a trait shared by Adam and Malcolm.

- Stuart

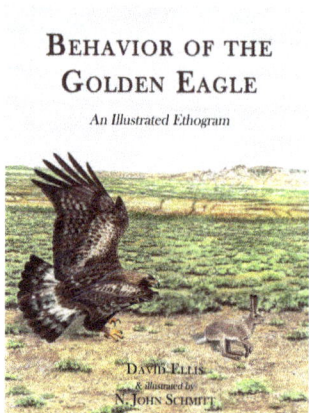
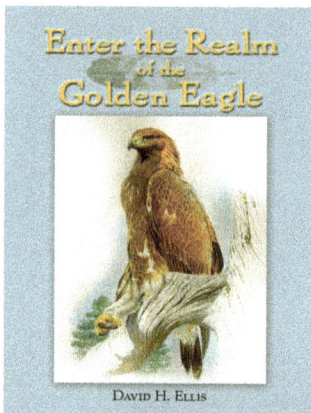